TEACHER'S EDITION

INVESTIGATING

AN INQUIRY EARTH SCIENCE PROGRAM

INVESTIGATING OUR DYNAMIC PLANET

Michael J. Smith Ph.D.
American Geological Institute

John B. Southard Ph.D.
Massachusetts Institute of Technology

Colin Mably
Curriculum Developer

Developed by the American Geological Institute
Supported by the National Science Foundation and
the American Geological Institute Foundation

Published by
It's About Time, Inc., Armonk, NY

It's About Time, Inc.

84 Business Park Drive, Armonk, NY 10504
Phone (914) 273-2233 Fax (914) 273-2227
Toll Free (888) 698-TIME
www.Its-About-Time.com

President
Laurie Kreindler

Project Editor
Ruta Demery

Contributing Writer
William Jones

Design
John Nordland

Production Manager
Joan Lee

Associate Editor
Al Mari

All student activities in this textbook have been designed to be as safe as possible, and have been reviewed by professionals specifically for that purpose. As well, appropriate warnings concerning potential safety hazards are included where applicable to particular activities. However, responsibility for safety remains with the student, the classroom teacher, the school principals, and the school board.

Investigating Earth Systems™ is a registered trademark of the American Geological Institute. Registered names and trademarks, etc., used in this publication, even without specific indication thereof, are not to be considered unprotected by law.

It's About Time® is a registered trademark of It's About Time, Inc. Registered names and trademarks, etc., used in this publication, even without specific indication thereof, are not to be considered unprotected by law.

Printed and bound in the United States of America

ISBN #1-58591-094-5

1 2 3 4 5 QC 06 05 04 03 02

This project was supported, in part, by the
National Science Foundation (grant no. 9353035)

Opinions expressed are those of the authors and not necessarily those of the National Science Foundation or the donors of the American Geological Institute Foundation.

Student's Edition Illustrations and Photos

P5, P13, P15, P18, P19, P20, P23, P24, P25, P27, P28, P31, P33, P36, P37, P38, P39, P43 (map), P45 (map), P52, P54 (source: Edwin Colbert, "Wandering Lands and Animals." Dover Publishers, 1985),

P55 (map), P57, P58, P67, illustrations by Stuart Armstrong;

P41, Eric Bergmanis;

P8, technical art by Burmar Technical Corporation;

P22, Bob Christman;

Pxi (lower left), Digital Vision Royalty Free Images: North America;

P30, Perle Dorr;

Pxii, P2, P9, P11, P33, P43, P53, P63, illustrations by Dennis Falcon;

P4, courtesy of Fisher Science Education Co.;

P61, Jim D. Griggs, U.S. Geological Survey Hawaiian Volcano Observatory;

P68, Roger Hutchison;

Pxi (top left), Pxii, P51, Martin Miller;

P1, courtesy of the Paleontological Research Institute;

P6, PhotoDisc;

P37, John Shelton; P46, P66, J.K. Nakata; Pxi (upper right), H.G. Wilshire: from the "The October 17, 1989, Loma Prieta, California, Earthquake – Selected Photographs," U.S. Geological Survey;

P49, Barbara Zahm

Taking Full Advantage of Investigating Earth Systems Through Professional Development

Implementing a new curriculum is challenging. That is why It's About Time, Inc. has partnered with the American Geological Institute, developers of *Investigating Earth Systems (IES)*, to provide a full range of professional development services. The sessions described below were designed to help you deepen your understanding of the content, pedagogy, and assessment strategies outlined in this Teacher's Edition, and adapt the program to suit the needs of your students and your local and state standards and curriculum frameworks.

Professional Development Services Available

Implementation Workshops

Two to five-day sessions held at your site that prepare you to implement the inquiry, systems, and community-based approach to learning Earth Science featured in *IES*. These workshops can be tailored to serve the needs of your school district, with chapters selected from the modules based on local or state curricula and framework criteria.

Program Overviews

One to three-day introductory sessions that provide a complete overview of the content and pedagogy of the *IES* program, as well as hands-on experience with activities from specific chapters. Program overviews are designed in consultation with school districts, counties, and SSI organizations.

Regional New-Teacher Summer Institutes

Two to five-day sessions that are designed to deepen your Earth Science content knowledge, and to prepare you to teach through inquiry. Guidance is provided in the gathering and use of appropriate materials and resources and specific attention is directed to the assessment of student learning.

Leadership Institutes

Six-day summer sessions conducted by the American Geological Institute that are designed to prepare current users for professional development leadership and mentoring within their districts or as consultants for It's About Time.

Follow-up Workshops

One to two-day sessions that provide additional Earth Science content and pedagogy support to teachers using the program. These workshops focus on identifying and solving practical issues and challenges to implementing an inquiry-based program.

Mentoring Visits

One-day visits that can be tailored to your specific needs that include class visits, mentoring teachers of the program, and in-service sessions.

Please fill in the form below to receive more information about participating in one of these Professional Development Services. The form can be directly faxed to our Professional Development at 914-273-2227. Our department will contact you to discuss further details and fees.

District/School: _____ Phone: _____

Address: _____

Contact Name: _____ Title: _____

E-mail: _____ Fax: _____

School Enrollment: _____ Number of Students Impacted: _____ Grade Level: _____

Have you purchased the following: ❏ Student Editions ❏ Teacher Editions ❏ Kits

Briefly explain how you plan to implement or how you are implementing the program in your school.

_____

_____

Table of Contents

Investigating Earth Systems Team

Project Staff

Michael J. Smith, Principal Investigator
    Director of Education, American Geological Institute
John B. Southard, Senior Writer
    Professor of Geology, Massachusetts Institute of Technology
William O. Jones, Contributing Writer
    American Geological Institute
Caitlin N. Callahan, Project Assistant
    American Geological Institute
William S. Houston, Field Test Coordinator
    American Geological Institute
Harvey Rosenbaum, Field Test Evaluator
    Montgomery County School District, Maryland
Fred Finley, Project Evaluator
    University of Minnesota
Lynn Lindow, Pilot Test Evaluator
    University of Minnesota

Original Project Personnel

Robert L. Heller, Principal Investigator
Charles Groat, United States Geological Survey
Colin Mably, LaPlata, Maryland
Robert Ridky, University of Maryland
Marilyn Suiter, American Geological Institute

National Advisory Board

Jane Crowder
      Middle School Teacher, WA
Kerry Davidson
      Louisiana Board of Regents, LA
Joseph D. Exline
      Educational Consultant, VA
Louis A. Fernandez
      California State University, CA
Frank Watt Ireton
      National Earth Science Teachers Association, DC
LeRoy Lee
      Wisconsin Academy of Sciences, Arts and Letters, WI
Donald W. Lewis
      Chevron Corporation, CA
James V. O'Connor (deceased)
      University of the District of Columbia, DC
Roger A. Pielke Sr.
      Colorado State University, CO
Dorothy Stout
      Cypress College, CA
Lois Veath
      Advisory Board Chairperson - Chadron State College, NE

National Science Foundation Program Officers

Gerhard Salinger
Patricia Morse

Acknowledgements

Principal Investigator

Michael Smith is Director of Education at the American Geological Institute in Alexandria, Virginia. Dr. Smith worked as an exploration geologist and hydrogeologist. He began his Earth Science teaching career with Shady Side Academy in Pittsburgh, PA in 1988 and most recently taught Earth Science at the Charter School of Wilmington, DE. He earned a doctorate from the University of Pittsburgh's Cognitive Studies in Education Program and joined the faculty of the University of Delaware School of Education in 1995. Dr. Smith received the Outstanding Earth Science Teacher Award for Pennsylvania from the National Association of Geoscience Teachers in 1991, served as Secretary of the National Earth Science Teachers Association, and is a reviewer for Science Education and The Journal of Research in Science Teaching. He worked on the Delaware Teacher Standards, Delaware Science Assessment, National Board of Teacher Certification, and AAAS Project 2061 Curriculum Evaluation programs.

Senior Writer

John Southard received his undergraduate degree from the Massachusetts Institute of Technology in 1960 and his doctorate in geology from Harvard University in 1966. After a National Science Foundation postdoctoral fellowship at the California Institute of Technology, he joined the faculty at the Massachusetts Institute of Technology, where he is currently Professor of Geology emeritus. He was awarded the MIT School of Science teaching prize in 1989 and was one of the first cohorts of first MacVicar Fellows at MIT, in recognition of excellence in undergraduate teaching. He has taught numerous undergraduate courses in introductory geology, sedimentary geology, field geology, and environmental Earth Science both at MIT and in Harvard's adult education program. He was editor of the Journal of Sedimentary Petrology from 1992 to 1996, and he continues to do technical editing of scientific books and papers for SEPM, a professional society for sedimentary geology. Dr. Southard received the 2001 Neil Miner Award from the National Association of Geoscience Teachers.

Project Director/Curriculum Designer

Colin Mably has been a key curriculum developer for several NSF-supported national curriculum projects. As learning materials designer to the American Geological Institute, he has directed the design and development of the IES curriculum modules and also training workshops for pilot and field-test teachers.

Project Team

Marcus Milling
Executive Director - AGI, VA

Michael Smith
Principal Investigator - Director of
Education - AGI, VA

Colin Mably
Project Director/Curriculum Designer -
Educational
Visions, MD

Matthew Smith
Project Coordinator
Program Manager - AGI, VA

Fred Finley
Project Evaluator
University of Minnesota, MN

Joe Moran
American Meteorological Society

Lynn Lindow
Pilot Test Evaluator
University of Minnesota, MN

Harvey Rosenbaum
Field Test Evaluator
Montgomery School
District, MD

Ann Benbow
Project Advisor - American Chemical
Society, DC

Robert Ridky
Original Project Director
University of Maryland, MD

Chip Groat
Original Principal Investigator -
University of Texas
El Paso, TX

Marilyn Suiter
Original Co-principal Investigator - AGI,
VA

William Houston
Field Test Manager

Caitlin Callahan - Project Assistant

**Original and
Contributing Authors**

Oceans
George Dawson
Florida State University, FL

Joseph F. Donoghue
Florida State University, FL

Ann Benbow
American Chemical Society

Michael Smith
American Geological Institute

Soil
Robert Ridky
University of Maryland, MD

Colin Mably - LaPlata, MD

John Southard
Massachusetts Institute of Technology,
MA

Michael Smith
American Geological Institute

Fossils
Robert Gastaldo
Colby College, ME

Colin Mably - LaPlata, MD

Michael Smith
American Geological Institute

Climate and Weather
Mike Mogil
How the Weather Works, MD

Ann Benbow
American Chemical Society

Joe Moran
American Meteorological Society

Michael Smith
American Geological Institute

Energy Resources
Laurie Martin-Vermilyea
American Geological Institute

Michael Smith
American Geological Institute

Dynamic Planet
Michael Smith
American Geological Institute

Rocks and Landforms
Michael Smith
American Geological Institute

Water as a Resource
Ann Benbow
American Chemical Society

Michael Smith
American Geological Institute

Materials and Minerals
Mary Poulton
University of Arizona, AZ

Colin Mably - LaPlata, MD

Michael Smith
American Geological Institute

Advisory Board

Jane Crowder
Middle School Teacher, WA

Kerry Davidson
Louisiana Board of Regents, LA

Joseph D. Exline
Educational Consultant, VA

Louis A. Fernandez
California State University, CA

Frank Watt Ireton
National Earth Science Teachers
Association, DC

LeRoy Lee
Wisconsin Academy of Sciences, Arts and
Letters, WI

Donald W. Lewis
Chevron Corporation, CA

James V. O'Connor (deceased)
University of the District of Columbia,
DC

Roger A. Pielke Sr.
Colorado State University, CO

Dorothy Stout
Cypress College, CA

Lois Veath
Advisory Board Chairperson
Chadron State College, NE

Pilot Test Teachers

Debbie Bambino
Philadelphia, PA

Barbara Barden - Rittman, OH

Louisa Bliss - Bethlehem, NH

Mike Bradshaw - Houston TX

Greta Branch - Reno, NV

Garnetta Chain - Piscataway, NJ

Roy Chambers Portland, OR

Laurie Corbett - Sayre, PA

James Cole - New York, NY

Collette Craig - Reno, NV

Anne Douglas - Houston, TX

Jacqueline Dubin - Roslyn, PA

Jane Evans - Media, PA

Gail Gant - Houston, TX

Joan Gentry - Houston, TX

Pat Gram - Aurora, OH

Robert Haffner - Akron, OH

Joe Hampel - Swarthmore, PA

Wayne Hayes - West Green, GA

Mark Johnson - Reno, NV

Cheryl Joloza - Philadelphia, PA

Jeff Luckey - Houston, TX

Karen Luniewski
Reistertown, MD

Cassie Major - Plainfield, VT

Carol Miller - Houston, TX

Melissa Murray - Reno, NV

Mary-Lou Northrop
North Kingstown, RI

Keith Olive - Ellensburg, WA

Tracey Oliver - Philadelphia, PA

Nicole Pfister - Londonderry, VT

Beth Price - Reno, NV

Joyce Ramig - Houston, TX

Julie Revilla - Woodbridge, VA

Steve Roberts - Meredith, NH

Cheryl Skipworth
Philadelphia, PA

Brent Stenson - Valdosta, GA

Elva Stout - Evans, GA

Regina Toscani
Philadelphia, PA

Bill Waterhouse
North Woodstock, NH

Leonard White
Philadelphia, PA

Paul Williams - Lowerford, VT

Bob Zafran - San Jose, CA

Missi Zender - Twinsburg, OH

Field Test Teachers

Eric Anderson - Carson City, NV
Katie Bauer - Rockport, ME
Kathleen Berdel - Philadelphia, PA
Wanda Blake - Macon, GA
Beverly Bowers
Mannington, WV
Rick Chiera - Monroe Falls, OH
Don Cole - Akron, OH
Patte Cotner - Bossier City, LA
Johnny DeFreese - Haughton, LA
Mary Devine - Astoria, NY
Cheryl Dodes - Queens, NY

Brenda Engstrom - Warwick, RI
Lisa Gioe-Cordi - Brooklyn, NY
Pat Gram - Aurora, OH
Mark Johnson - Reno, NV
Chicory Koren - Kent, OH
Marilyn Krupnick
Philadelphia, PA
Melissa Loftin - Bossier City, LA
Janet Lundy - Reno, NV
Vaughn Martin - Easton, ME
Anita Mathis - Fort Valley, GA
Laurie Newton - Truckee, NV
Debbie O'Gorman - Reno, NV

Joe Parlier - Barnesville, GA
Sunny Posey - Bossier City, LA
Beth Price - Reno, NV
Stan Robinson
Mannington, WV
Mandy Thorne
Mannington, WV
Marti Tomko
Westminster, MD
Jim Trogden - Rittman, OH
Torri Weed - Stonington, ME
Gene Winegart - Shreveport, LA
Dawn Wise - Peru, ME
Paula Wright - Gray, GA

IMPORTANT NOTICE

The *Investigating Earth Systems*™ series of modules is intended for use by students under the direct supervision of a qualified teacher. The experiments described in this book involve substances that may be harmful if they are misused or if the procedures described are not followed. Read cautions carefully and follow all directions. Do not use or combine any substances or materials not specifically called for in carrying out experiments. Other substances are mentioned for educational purposes only and should not be used by students unless the instructions specifically indicate.

The materials, safety information, and procedures contained in this book are believed to be reliable. This information and these procedures should serve only as a starting point for classroom or laboratory practices, and they do not purport to specify minimal legal standards or to represent the policy of the American Geological Institute. No warranty, guarantee, or representation is made by the American Geological Institute as to the accuracy or specificity of the information contained herein, and the American Geological Institute assumes no responsibility in connection therewith. The added safety information is intended to provide basic guidelines for safe practices. It cannot be assumed that all necessary warnings and precautionary measures are contained in the printed material and that other additional information and measures may not be required.

This work is based upon work supported by the National Science Foundation under Grant No. 9353035 with additional support from the Chevron Corporation. Any opinions, findings, and conclusions or recommendations expressed in this publication are those of the authors and do not necessarily reflect the views of the National Science Foundation or the Chevron Corporation. Any mention of trade names does not imply endorsement from the National Science Foundation or the Chevron Corporation.

The American Geological Institute and Investigating Earth Systems

Imagine more than 500,000 Earth scientists worldwide sharing a common voice, and you've just imagined the mission of the American Geological Institute. Our mission is to raise public awareness of the Earth sciences and the role that they play in mankind's use of natural resources, mitigation of natural hazards, and stewardship of the environment. For more than 50 years, AGI has served the scientists and teachers of its Member Societies and hundreds of associated colleges, universities, and corporations by producing Earth science educational materials, *Geotimes*–a geoscience news magazine, GeoRef–a reference database, and government affairs and public awareness programs.

So many important decisions made every day that affect our lives depend upon an understanding of how our Earth works. That's why AGI created *Investigating Earth Systems*. In your *Investigating Earth Systems* classroom, you'll discover the wonder and importance of Earth science. As you investigate minerals, soil, or oceans — do field work in nearby beaches, parks, or streams, explore how fossils form, understand where your energy resources come from, or find out how to forecast weather — you'll gain a better understanding of Earth science and its importance in your life.

We would like to thank the National Science Foundation and the AGI Foundation Members that have been supportive in bringing Earth science to students. The Chevron Corporation provided the initial leadership grant, with additional contributions from the following AGI Foundation Members: Anadarko Petroleum Corp., The Anschutz Foundation, Baker Hughes Foundation, Barrett Resources Corp., Elizabeth and Stephen Bechtel, Jr. Foundation, BPAmoco Foundation, Burlington Resources Foundation, CGG Americas, Inc., Conoco Inc., Consolidated Natural Gas Foundation, Diamond Offshore Co., Dominion Exploration & Production, Inc., EEX Corp., ExxonMobil Foundation, Global Marine Drilling Co., Halliburton Foundation, Inc., Kerr McGee Foundation, Maxus Energy Corp., Noble Drilling Corp., Occidental Petroleum Charitable Foundation, Parker Drilling Co., Phillips Petroleum Co., Santa Fe Snyder Corp., Schlumberger Foundation, Shell Oil Company Foundation, Southwestern Energy Co., Texaco, Inc., Texas Crude Energy, Inc., Unocal Corp. USX Foundation (Marathon Oil Co.).

We at AGI wish you success in your exploration of the Earth System!

Michael J. Smith
Director of Education, AGI

Marcus E. Milling
Executive Director, AGI

Developing Investigating Earth Systems

Welcome to *Investigating Earth Systems (IES)!* *IES* was developed through funding from the National Science Foundation and the American Geological Institute Foundation. Classroom teachers, scientists, and thousands of students across America helped to create *IES*. In the 1997-98 school year, scientists and curriculum developers drafted nine *IES* modules. They were pilot tested by 43 teachers in 14 states from Washington to Georgia. Faculty from the University of Minnesota conducted an independent evaluation of the pilot test in 1998, which was used to revise the program for a nationwide field test during the 1999-2000 school year. A comprehensive evaluation of student learning by a professional field-test evaluator showed that *IES* modules led to significant gains in student understanding of fundamental Earth science concepts. Field-test feedback from 34 teachers and content reviews from 33 professional Earth scientists were used to produce the commercial edition you have selected for your classroom.

Inquiry and the interrelation of Earth's systems form the backbone of *IES*. Often taught as a linear sequence of events called "the scientific method," inquiry underlies all scientific processes and can take many different forms. It is very important that students develop an understanding of inquiry processes as they use them. Your students naturally use inquiry processes when they solve problems. Like scientists, students usually form a question to investigate after first looking at what is observable or known. They predict the most likely answer to a question. They base this prediction on what they already know to be true. Unlike professional scientists, your students may not devote much thought to these processes. In order to be objective, students must formally recognize these processes as they do them. To make sure that the way they test ideas is fair, scientists think very carefully about the design of their investigations. This is a skill your students will practice throughout each *IES* module.

All *Investigating Earth Systems* modules also encourage students to think about the Earth as a system. Upon completing each investigation they are asked to relate what they have learned to the Earth Systems (see the *Earth System Connection* sheet in the **Appendix**). Integrating the processes of the biosphere, geosphere, hydrosphere, and atmosphere will open up a new way of looking at the world for most students. Understanding that the Earth is dynamic and that it affects living things, often in unexpected ways, will engage them and make the topics more relevant.

We trust that you will find the Teacher's Edition that accompanies each student module to be useful. It provides **Background Information** on the concepts explored in the module, as well as strategies for incorporating inquiry and a systems-based approach into your classroom. Enjoy your investigation!

Investigating Earth Systems Modules

Climate and Weather

Our Dynamic Planet

Energy Resources

Fossils

Materials and Minerals

Oceans

Rocks and Landforms

Soil

Water as a Resource

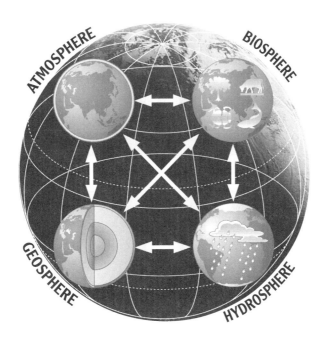

Investigating Earth Systems: Correlation to the National Science Education Standards

National Science Education Content Standards Grades 5 – 8	Soil	Rocks and Landforms	Oceans	Climate and Weather	Our Dynamic Planet	Materials and Minerals	Energy Resources	Water as a Resource	Fossils
UNIFYING CONCEPTS AND PROCESSES									
System, order and organization	•	•	•	•	•	•	•	•	•
Evidence, models, and explanation	•	•	•	•	•	•	•	•	•
Constancy, change, and measurement	•	•	•	•	•	•	•	•	•
Evolution and equilibrium		•	•	•	•			•	•
Form and function									•
SCIENCE AS INQUIRY									
Identify questions that can be answered through scientific investigations	•	•	•	•	•	•	•	•	•
Design and conduct scientific investigations	•	•	•	•	•	•	•	•	•
Use tools and techniques to gather, analyze, and interpret data	•	•	•	•	•	•	•	•	•
Develop descriptions, explanations, predictions and models based on evidence	•	•	•	•	•	•	•	•	•
Think critically and logically to make the relationships between evidence and explanation	•	•	•	•	•	•	•	•	•
Recognize and analyze alternative explanations and predictions	•	•	•	•	•	•	•	•	•
Communicate scientific procedures and explanations	•	•	•	•	•	•	•	•	•
Use mathematics in all aspects of scientific inquiry	•	•	•	•	•	•	•	•	•
Understand scientific inquiry	•	•	•	•	•	•	•	•	•
PHYSICAL SCIENCE									
Properties and Changes of Properties in Matter	•	•	•		•	•	•	•	
Motions and Forces	•		•						
Transfer of Energy		•	•	•	•	•	•		
LIFE SCIENCE									
Populations and Ecosystems			•				•	•	•
Diversity and Adaptation of Organisms			•		•				•

Investigating Earth Systems: Correlation to the National Science Education Standards

National Science Education Content Standards Grades 5 – 8

	Soil	Rocks and Landforms	Oceans	Climate and Weather	Our Dynamic Planet	Materials and Minerals	Energy Resources	Water as a Resource	Fossils
EARTH AND SPACE SCIENCE									
Structure of the Earth system	•	•	•	•	•	•	•	•	•
Earth's History	•	•	•	•	•	•	•	•	•
Earth in the Solar System			•	•	•		•	•	
SCIENCE AND TECHNOLOGY									
Abilities of technological design	•	•	•	•	•	•	•	•	•
Understandings about science and technology		•	•			•	•	•	
SCIENCE IN PERSONAL AND SOCIAL PERSPECTIVES									
Personal health	•							•	
Populations, resources, and environment	•					•	•	•	
Natural Hazards		•		•	•	•			
Risks and benefits					•		•		
Science and technology in society	•	•	•	•	•	•	•	•	•
HISTORY AND NATURE OF SCIENCE									
Science as a human endeavor	•	•	•	•	•	•	•	•	•
Nature of science	•	•	•	•	•	•	•	•	•
History of science			•		•				•

Using Investigating Earth Systems Features in Your Classroom

1. Pre-assessment

Designed under the umbrella framework of "science for all students," meaning that all students should be able to engage in inquiry and learn core science concepts, *Investigating Earth Systems* helps you to tailor instruction to meet your students' needs. A crucial first step in this framework is to ascertain what knowledge, experience, and understanding your students bring to their study of a module. The pre-assessment consists of four questions geared to the major concepts and understandings targeted in the unit. Students write and draw what they know about the major topics and concepts. This information is recorded and shared in an informal discussion prior to engaging in hands-on inquiry. The discussion enables students to recognize how much there is to learn and appreciate, and that by exploring the unit together, the entire classroom can emerge from the experience with a better understanding of core concepts and themes. Students' responses provide crucial pre-assessment data for you. By examining their written work and probing for further detail during the classroom conversation, you can identify strengths and weaknesses in students' understandings, as well as their abilities to communicate that understanding to others. It is important that the pre-assessment not be viewed as a test, and that judgments about the accuracy of responses be evaluated in writing or through your comments during the conversation. The goal is to ascertain and probe, not judge, and to create a safe classroom environment in which students feel comfortable sharing their ideas and knowledge. Students revisit these pre-assessment questions informally throughout the unit. At the end of the unit, students respond to the same four questions in the section called **Back to the Beginning**. The pre-assessment thus helps you and your students to make judgments about their growth in understanding and ability throughout the module.

2. The Earth System

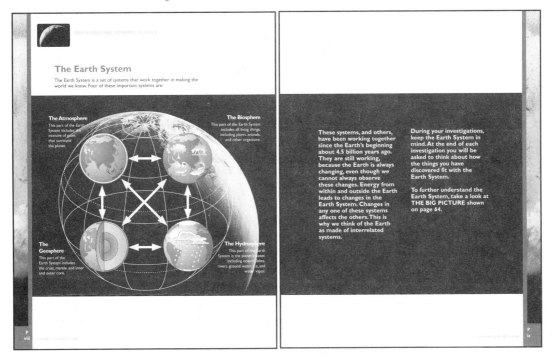

National Science Education Standards link...

"A major goal of science in the middle grades is for students to develop an understanding of Earth (and the solar system) as a set of closely coupled systems. The idea of systems provides a framework in which students can investigate the four major interacting components of the Earth System – geosphere (crust, mantle, and core), hydrosphere (water), atmosphere (air), and the biosphere (the realm of living things)."

NSES content standard D "Developing Student Understanding" (pages 158-159)

Understanding the Earth system is an overall goal of the *Investigating Earth Systems* series. It is a difficult and complex set of concepts to grasp, because it is inferred rather than observed directly. Yet even the smallest component of Earth science can be linked to the Earth system. As your students progress through each module, an increasing number of connections with the Earth system will arise. Your students may not, however, immediately see these connections. At the end of every investigation, they will be asked to link what they have discovered with ideas about the Earth system. They will also be asked to write about this in their journals. A **Blackline Master** (*Earth System Connection* sheet) is available in each Teacher's Edition. Students can use this to record connections that they make as they complete each investigation. At the very end of the module they will be asked to review everything they have learned in relation to the Earth system. The aim is for students to have a working understanding of the Earth System by the time they complete grade 8. They will need your help accomplishing this.

For example, in *Investigating Rocks and Landforms*, students work with models to simulate Earth processes, such as erosion of stream sediment and deposition of that sediment on floodplains and in deltas. Changes in inputs in one part of the system (say rainfall, from the atmosphere), affect other parts of the system (stream flows, erosion on river bends, amount of sediment carried by the stream, and deposition of sediment on floodplains or in deltas). These changes affect, in turn, other parts of the system (for example, floods that affect human populations, i.e., the biosphere). In the same module, students explore the rock record within their community and develop understandings about how interactions between the hydrosphere, atmosphere, geosphere, and biosphere change the landscape over time. These are just some of the many ways that *Investigating Earth Systems* modules foster and promote student thinking about the dynamic nature and interactions of Earth systems—biosphere, geosphere, atmosphere, and hydrosphere.

3. Introducing Inquiry Processes

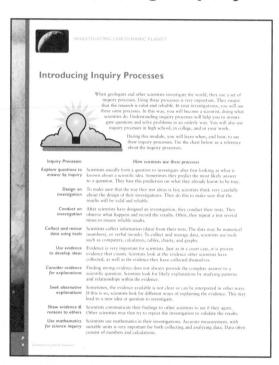

Inquiry is at the heart of *Investigating Earth Systems*. That is why each module title begins with the title "Investigating." In the National Science Education Standards, inquiry is the first Content Standard. NSES then lists a range of points about inquiry. These fundamental components of inquiry were written into the list shown at the beginning of each student module. It is very important that students be reminded of the steps in the inquiry process as they perform them. Inquiry depends on active student participation. Ideas on how to make inquiry successful in the classroom appear throughout the modules and in the "Managing Inquiry in Your *Investigating Earth Systems* Classroom" section of this Teacher's Edition.

It is very important that students develop an understanding of the inquiry processes as they use them. Stress the importance of inquiry processes as they occur in your investigations. Provoke students to think about why these processes are important. Collecting good data, using evidence, considering alternative explanations, showing evidence to others, and using mathematics are all essential to *IES*. Use examples to demonstrate these processes whenever possible. At the end of every investigation, students are asked to reflect on the scientific inquiry processes they used. Refer students to the list of inquiry processes on page x of the Student Book as they think about scientific inquiry and answer the questions.

4. Introducing the Module

Each *IES* module begins with photographs and questions. This is an introduction to the module for your students. It is designed to give them a brief overview of the content of the module and set their investigations into a relevant and meaningful context. Students will have had a variety of experiences with the content of the module. This is an opportunity for them to offer some of their own experiences in a general discussion, using these questions as prompts. This section of each *IES* module follows the pre-assessment, where students spend time thinking about what they already know about the content of the module. The photographs and questions can be used to focus the students' thinking.

The ideas students share in the introduction to the module provide you with additional pre-assessment data. The experiences they describe and the way in which they are discussed will alert you to their general level of understanding about these topics. To encourage sharing and to provide a record, teachers find it useful to quickly summarize the main points that emerge from discussion. You can do this on the chalkboard or flipchart for all to see. This can be displayed as students work through the module and added to with each new experience. For your own assessment purposes, it will be useful to keep a record of these early indicators of student understanding.

5. Key Question

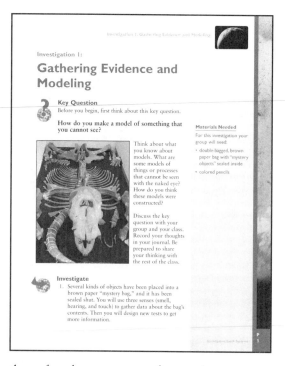

Each *Investigating Earth Systems* investigation begins with a **Key Question** – an open-ended question that gives teachers the opportunity to explore what their students know about the central concepts of the activity. Uncovering students' thinking (their prior knowledge) and exposing the diversity of ideas in the classroom are the first steps in the learning cycle. One of the most fundamental principles derived from many years of research on student learning is that:

"Students come to the classroom with preconceptions about how the world works. If their initial understanding is not engaged, they may fail to grasp the new concepts and information that are taught, or they may learn them for the purposes of a test but revert to their preconceptions outside the classroom." (*How People Learn: Bridging Research and Practice*, National Research Council, 1999, P. 10.)

This principle has been illustrated through the *Private Universe* series of videotapes that show Harvard graduates responding to basic science questions in much the same way that fourth grade students do. Although the videotapes revealed that the Harvard graduates used a more sophisticated vocabulary, the majority held onto the same naïve, incorrect conceptions of elementary school students. Research on learning suggests that the belief systems of students who are not confronted with what they believe and adequately shown why they should give up that belief system remain intact. Real learning requires confronting one's beliefs and testing them in light of competing explanations.

Drawing out and working with students' preconceptions is important for learners. In *Investigating Earth Systems*, the **Key Question** is used to ascertain students' prior knowledge about the key concept or Earth science processes or events explored in the activity. Students verbalize what they think about the age of the Earth, the causes of volcanoes, or the way that the landscape changes over time before they embark on an activity designed to challenge and test these beliefs. A brief discussion about the diversity of beliefs in the classroom makes students consider how their ideas compare to others and the evidence that supports their view of volcanoes, earthquakes, or seasons.

The **Key Question** is not a conclusion, but a lead into inquiry. It is not designed to instantly yield the "correct answer" or a debate about the features of the question, or to bring closure. The activity that follows will provide that discussion as students analyze and discuss the results of inquiry. Students are encouraged to record their ideas in words and/or drawings to ensure that they have considered their prior knowledge. After students discuss their ideas in pairs or in small groups, teachers activate a class discussion. A discussion with fellow students prior to class discussion may encourage students to exchange ideas without the fear of personally giving a "wrong answer." Teachers sometimes have students exchange papers and volunteer responses that they find interesting.

Some teachers prefer to have students record their responses to these questions. They then call for volunteers to offer ideas up for discussion. Other teachers prefer to start with discussion by asking students to volunteer their ideas. In either situation, it is important that teachers encourage the sharing of ideas by not judging responses as "right" or "wrong." It is also important that teachers keep a record of the variety of ideas, which can be displayed in the classroom (on a sheet of easel pad paper or on an overhead transparency) and referred to as students explore the concepts in the module. Teachers often find that they can group responses into a few categories and record the number of students who hold each idea. The photograph in each **Key Question** section was designed to stimulate student thinking and help students to make the specific kinds of connections emphasized in each activity.

6. Investigate

Investigating Earth Systems is a hands-on, minds-on curriculum. In designing *Investigating Earth Systems*, we were guided by the research on learning, which points out how important *doing* Earth Science is to *learning* Earth Science. Testing of *Investigating Earth Systems* activities by teachers across America provided critical testimonial and quantitative measures of the importance of the activities to student learning. In small groups and as a class, students take part in doing hands-on experiments, participating in field work, or searching for answers using the Internet and reference materials. **Blackline Masters** are included in the Teacher's Editions for any maps or illustrations that are essential for students to complete the activity.

Each part of an *Investigating Earth Systems* investigation, as well as the sequence of activities within a module, moves from concrete to abstract. Hands-on activities provide the basis for exploring student beliefs about how the world works and to manipulate variables that affect the outcomes of experiments, models, or simulations. Later in each activity, formal labels are applied to concepts by introducing terminology used to describe the processes that students have explored through hands-on activity. This flow from concrete (hands-on) to abstract (formal explanations) is progressive – students begin to develop their own explanations for phenomena by responding to questions within the **Investigate** section.

Each activity has instructions for each part of the investigation. Materials kits are available for purchase, but you will also need to obtain some resources from outside suppliers, such as topographic and geologic maps of your community, state, or region. The *Investigating Earth Systems* web site will direct you to sources where you can gather such materials.

Most **Investigate** activities will require between one and two class periods. The variety of school schedules and student needs makes it difficult to predict exactly how much time your class will need. For example, if students need to construct a graph for part of an investigation, and the students have never been exposed to graphing, then this investigation may require additional time and could become part of a mathematics lesson.

The most challenging aspect of *Investigating Earth Systems* for teachers to "master" is that the **Investigate** section of each activity has been designed to be student-driven. Students learn more when they have to struggle to "figure things out" and work in collaborative groups to solve problems as a team. Teachers will have to resist the temptation to provide the answers to students when they get "stuck" or hung up on part of a problem. Eventually, students learn that while they can call upon their teacher for assistance, the teacher is not going to "show them the answer." Field testing of *Investigating Earth Systems* revealed that teachers who were most successful in getting their students to solve problems as a team were patient with this process and steadfast in their determination to act as facilitators of learning during the **Investigate** portion of activities. As one teacher noted, "My response to questions during the investigation was like a mantra, 'What do you think you need to do to solve this?' My students eventually realized that although I was there to provide guidance, they weren't going to get the solution out of me."

Another concern that many teachers have when examining *Investigating Earth Systems* for the first time is that their students do not have the background knowledge to do the investigations. They want to deliver a lecture about the phenomena before allowing students to do the investigation. Such an approach is common to many traditional programs and is inconsistent with the pedagogical theory used to design *Investigating Earth Systems*. The appropriate place for delivering a lecture or reading text in *Investigating Earth Systems* is following the investigation, not preceding it.

For example, suppose a group of students has been asked to interpret a map. The traditional approach to science education is for the teacher to give a lecture or assign a reading, "How to Interpret Maps," then give students practice reading maps. *Investigating Earth Systems* teachers recognize that while students may lack some specific skills (reading latitude and longitude, for example), within a group of four students, it is not uncommon for at least one of the students to have a vital skill or piece of knowledge that is required to solve a problem. The one or two students who have been exposed to (or better yet, understood) latitude and longitude have the opportunity to shine within the group by contributing that vital piece of information or demonstrating a skill. That's how scientific research teams work – specialists bring expertise to the group, and by working together, the group achieves something that no one could achieve working alone. The **Investigate** section of *Investigating Earth Systems* is modeled in the spirit of the scientific research team.

7. Inquiry

Inquiry is the first content standard in the National Science Education Standards (NSES). The American Association for the Advancement of Science's (AAAS) Benchmarks for Science Literacy also places considerable emphasis on scientific inquiry (see excerpts on the following page). *IES* has been designed to remind students to reflect on inquiry processes as they carry out their investigations. The student journal is an important tool in helping students to develop these understandings. In using the journal, students are modeling what scientists do. Your students are young scientists as they investigate Earth science questions. Encourage your students to think of themselves in this way and to see their journals as records of their investigations.

Inquiry

Representing Information

Communicating findings to other scientists is very important in scientific inquiry. In this investigation it is important for you to find good ways of showing what you learned to others in your class. Be sure your maps and displays are clearly labeled and well organized.

An icon was developed to draw students' attention to brief descriptions of inquiry processes in the margins of the student module. The icon and explanations provide opportunities to direct students' attention to what they are doing, and thus serve as an important metacognitive tool to stimulate thinking about thinking.

National Science Education Standards link...

<u>Content Standard A</u>
As a result of activities in grades 5-8, all students should develop:
- Abilities necessary to do scientific inquiry
- Understandings about scientific inquiry

<u>Abilities Necessary to do Scientific Inquiry</u>
- Identify questions that can be answered through scientific investigations
- Use appropriate tools and techniques to gather, analyze, and interpret data
- Develop descriptions, explanations, predictions, and models using evidence
- Think critically and logically to make the relationships between evidence and explanations
- Recognize and analyze alternative explanations and predictions
- Communicate scientific procedures and explanations
- Use mathematics in all aspects of scientific inquiry

(From National Science Education Standards, pages 145-148)

Benchmarks for Science Literacy link...

<u>The Nature of Science Inquiry: Grades 6 through 8</u>
- At this level, students need to become more systematic and sophisticated in conducting their investigations, some of which may last for several weeks. That means closing in on an understanding of what constitutes a good experiment. The concept of controlling variables is straightforward, but achieving it in practice is difficult. Students can make some headway, however, by participating in enough experimental investigations (not to the exclusion, of course, of other kinds of investigations) and explicitly discussing how explanation relates to experimental design.

- Student investigations ought to constitute a significant part—but only a part—of the total science experience. Systematic learning of science concepts must also have a place in the curriculum, for it is not possible for students to discover all the concepts they need to learn, or to observe all of the phenomena they need to encounter, solely through their own laboratory investigations. And even though the main purpose of student investigations is to help students learn how science works, it is important to back up such experience with selected readings. This level is a good time to introduce stories (true and fictional) of scientists making discoveries – not just world-famous scientists, but scientists of very different backgrounds, ages, cultures, places, and times.

(From Benchmarks for Science Literacy, page 12)

8. Digging Deeper

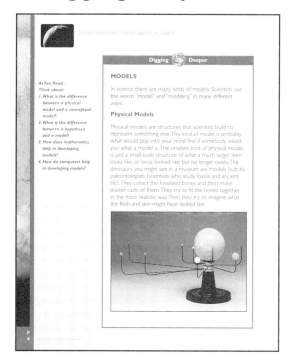

This section provides text, illustrations, data tables, and photographs that give students greater insight into the concepts explored in the activity. Teachers often assign **As You Read** questions as homework to guide students to think about the major ideas in the text. Teachers can also select questions to use as quizzes, rephrasing the questions into multiple choice or "true/false" formats. This provides assessment information about student understanding and serves as a motivational tool to ensure that students complete the reading assignment and comprehend the main ideas.

This is the stage of the activity that is most appropriate for teachers to explain concepts to students in whole-class lectures or discussions. References to **Blackline Masters** are available throughout the Teacher's Edition. They refer to illustrations from the textbook that teachers may photocopy and distribute to students or make overhead transparencies for lectures or presentations.

9. Review and Reflect

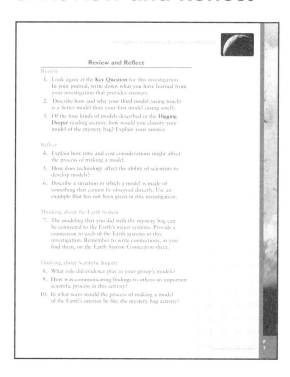

Questions in this feature ask students to use the key principles and concepts introduced in the activity. Students are sometimes presented with new situations in which they are asked to apply what they have learned. The questions in this section typically require higher-order thinking and reasoning skills than the **As You Read** questions. Teachers can assign these questions as homework, or have students complete them in groups during class. Assigning them as homework economizes time available in class, but has the drawback of making it difficult for students to collectively revisit the understanding that they developed as they worked through the concepts as a group during the investigation. A third alternative is, of course, to assign the work individually in class. When students work through application problems in class, teachers have the opportunity to interact with students at a critical juncture in their learning – when they may be just on the verge of "getting it."

Review and Reflect prompts students to think about what they have learned, how their work connects with the Earth system, and what they know about scientific inquiry. Another one of the important principles of learning used to guide the selection of content in *Investigating Earth Systems* was that:

"To develop competence in an area of inquiry, students must (a) have a deep foundation of factual knowledge, (b) understand facts and ideas in the context of a conceptual framework, and (c) organize knowledge in ways that facilitate retrieval and application." (*How People Learn: Bridging Research and Practice* National Research Council, 1999, P. 12.)

Reflecting on one's learning and one's thinking is an important metacognitive tool that makes students examine what they have learned in the activity and then think critically about the usefulness of the results of their inquiry. It requires students to take stock of their learning and evaluate whether or not they really understand "how it fits into the Big Picture." It is important for teachers to guide students through this process with questions such as "What part of your work demonstrates that you know and can do scientific inquiry? How does what you learned help you to better understand the Earth system? How does your work contribute or relate to the concepts of the Big Picture at the end of the module?"

10. Final Investigation: Putting It All Together

In the final investigation in each *Investigating Earth Systems* module, your students will apply all the knowledge they have about the topics explored to solve a practical problem or situation. Requiring students to apply all they have gained toward a specific outcome should serve as the main assessment information for the module. A sample assessment rubric is provided in the back of this Teacher's Edition. Whatever rubric you employ, it is important that you share this with students at the outset of the final investigation so that they understand the criteria upon which their work will be judged.

The instructions provided to students are purposely open-ended, but can be completed to various levels depending upon how much knowledge students apply. During the final investigation, your role is to be a participant observer, moving from group to group, noticing how students go about the investigation and how they are applying the experience and understanding they have gained from the module.

11. Reflecting

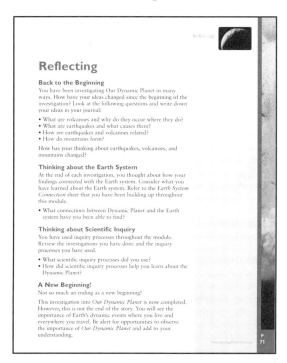

Now that students are at the end of the module, they are provided with questions that ask them to reflect upon all that they have learned about Earth science, inquiry, and the Earth system. The first set of questions (**Back to the Beginning**) are the same questions used in the pre-assessment. Teachers often ask students to revisit their initial responses and provide new answers to demonstrate how much they have learned.

12. The Big Picture

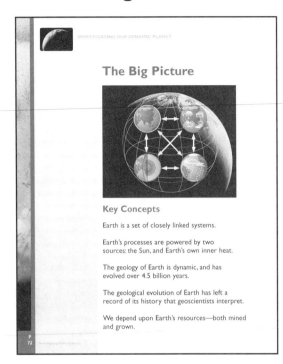

The five key concepts below underlie Earth science in general and *Investigating Earth Systems* in particular. Collectively, the nine modules in the *Investigating Earth Systems* series are designed to help students understand each of these concepts by the time they complete grade 8. Many of the concepts that underlie the Big Picture may be difficult for students to grasp easily. As students develop their ideas through inquiry-based investigations, you can help them to make connections with these key scientific concepts. As a reminder of the importance of the major understandings, the Student Book has a copy of the Big Picture in the back of the book near the **Glossary**.

Be on the lookout for chances to remind students that:
- Earth is a set of closely linked systems. *see pg*
- Earth's processes are powered by two sources: the Sun and Earth's own inner heat.
- The geology of Earth is dynamic, and has evolved over 4.5 billion years.
- The geological evolution of Earth has left a record of its history that geoscientists interpret.
- We depend upon Earth's resources—both mined and grown.

13. Glossary

Words that may be new or unfamiliar to students are defined and explained in the **Glossary** of the Student Book. Teachers use their own judgment about selecting the terms that appear in the **Glossary** that are most important for their students to learn. Teachers typically use discretion and consider their state and local guidelines for science content understanding when assigning importance to particular vocabulary, which in most cases is very likely to be a small subset of all the scientific terms introduced in each module and defined in the **Glossary**.

References

How People Learn: Bridging Research and Practice (1999) Suzanne Donovan, John Bransford, and James Pellegrino, editors. National Academy Press, Washington, DC. 78 pages. The report is also available online at www.nap.edu.

Using the Investigating Earth Systems Web Site

www.agiweb.org/ies

The *Investigating Earth Systems* web site has been designed for teachers and students.
- Each *Investigating Earth Systems* module has its own web page that has been designed specifically for the content addressed within that module.
- Module web sites are broken down by investigation and also contain a section with links to relevant resources that are useful for the module.
- Each investigation is divided into materials and supplies needed, **Background Information,** and links to resources that will help you and your students to complete the investigation.

Enhancing Teacher Content Knowledge

Each *Investigating Earth Systems* module has a specific web page that will help teachers to gather further **Background Information** about the major topics covered in each activity.

Example from *Investigating Rocks and Landforms* – Investigation 1

Different Types of Rock

To learn more about different types of rocks, visit the following web sites:

• What are the basic types of rock?, Rogue Community College
This site lists the basic descriptions of sedimentary, metamorphic and igneous rocks. Detailed information on each type of rock is also available.
(http://www.jersey.uoregon.edu/~mstrick/AskGeoMan/geoQuerry13.html)

1. *Sedimentary Rocks:*
• Sedimentary Rocks, University of Houston
Detailed description of the composition, classification, and formation of sedimentary rocks.
(http://ucaswww.mcm.uc.edu/geology/maynard/INTERNETGUIDE/appendf.htm)
• Image Gallery for Geology, University of North Carolina
See more examples of sedimentary rocks.
(http://www.geosci.unc.edu/faculty/glazner/Images/SedRocks/SedRocks.html)
• Sedimentary Rocks Laboratory, Georgia Perimeter College
Read a thorough discussion of clastic, chemical, and organic sedimentary rocks. Illustrations accompany each description.
(http://www.gpc.peachnet.edu/~pgore/geology/historical_lab/sedrockslab.php)
• Textures and Structures of Sedimentary Rocks, Duke University
View a collection of slides of different sedimentary rocks as either outcrops or thin sections viewed through a microscope.
(http://www.geo.duke.edu/geo41/seds.htm)
• Sedimentary Rocks, Washington State University
Learn more about sedimentary processes, environments of deposition in relation to different sedimentary rocks. Topics covered include depositional environments, chemical or mechanical weathering, deposition and lithification, and classification.
(http://www.wsu.edu/~geology/geol101/sedimentary/seds.htm)

Obtaining Resources

The inquiry focus of *Investigating Earth Systems* will require teachers to obtain local or regional maps, rocks, and data. The *Investigating Earth Systems* web site helps teachers to find such materials. The web page for each *Investigating Earth Systems* module provides a list of relevant web sites, maps, videos, books, and magazines. Specific links to sources of these materials are often provided.

Managing Inquiry in Your Investigating Earth Systems Classroom

Materials

The proper management of materials can make the difference between a productive, positive investigation and a frustrating one. If your school has purchased the materials kit (available through It's About Time Publishing) most materials have been supplied. In many cases there will be additional items that you will need to supply as well. This can include photocopies or transparencies (**Blackline Masters** are available in the **Appendix**), or basic classroom supplies like an overhead projector or water source. On occasion, students will bring in materials. If you do not have the materials kit, a master list of materials for the entire module precedes the first investigation. Tips on using and managing materials accompany each investigation.

Safety

Each activity has icons noting safety concerns. In most cases, a well-managed classroom is the best preventive measure for avoiding danger and injury. Take time to explain your expectations before beginning the first investigation. Read through the investigations with your students and note any safety concerns. The activities were designed with safety in mind and have been tested in classrooms. Nevertheless, be alert and observant at all times. Often, the difference between an accident and a calamity is simple monitoring.

Time

This module can be completed in six weeks if you teach science in daily 45-minute class periods. However, there are many opportunities to extend the investigations, and perhaps to shorten others. The nature of the investigations allows for some flexibility.

An inquiry approach to science education requires the careful management of time for students to fully develop their investigative experience and skills. Most investigations will not easily fit into one 45-minute lesson. You may feel it necessary to extend them over two or more class periods. Some investigations include long-term studies. Where this is the case you may need to allow time for data collection each day, even after moving on to the next investigation.

Classroom Space

On days when students work as groups, arrange your classroom furniture into small group areas. You may want to have two desk arrangements—one for group work and one for direct instruction or quiet work time.

The Student Journal

The student journal is an important component of each *IES* module. (See the **Appendix** in this Teacher's Edition for a **Blackline Master** of the Journal cover sheet.) Your students are young scientists as they investigate Earth science questions. Encourage your students to think of themselves in this way and to see their journals as records of their investigations.

The journal serves other functions as well. It is a key component in performance assessment, both formative and summative. (Formative evaluation involves the ongoing evaluation of students' level of understanding and their development of skills and attitudes. Summative evaluation is designed to determine the extent to which instructional objectives have been achieved for a topic.) Encourage your students to record observations, data, and experimental results in their journals. Answers to **Review and Reflect** questions at the end of each investigation should also be recorded in the journal. It is very important that students have enough time to review, reflect, and update their journals at the end of each investigation.

Frequent feedback is essential if students are to maintain good journals. This is difficult but not impossible. For many teachers, the prospect of collecting and assessing anywhere from 20 to over 100 journals in a planning period, then returning them the next day, seems prohibitive. This does not need to be the case. If you use a simple rubric, and collect journals often, it is possible to assess 100 journals in an hour. It may not be necessary to write comments every time you collect journals; in some cases, it is equally effective to address trends in student work in front of the whole class. For example, students will inevitably turn in journals that contain no dates and/or headings. This leaves many questions unanswered and makes their work very hard to interpret. You might want to consider keeping your own teacher journal for this module. This makes a great template for evaluating student journals. In addition to documenting class activities, you might want to make notes on classroom management strategies, materials and supplies, and procedural modifications. Sample rubrics are included in the **Appendix**.

Student Collaboration

The National Science Education Standards and Benchmarks for Science Literacy emphasize the importance of student collaboration. Scientists and others frequently work in teams to investigate questions and solve problems. There are times, however, when it is important to work alone. You may have students who are more comfortable working this way. Traditionally, the competitive nature of school curricula has emphasized individual effort through grading, "honors" classes, and so on. Many parents will have been through this experience themselves as students and will be looking for comparisons between their children's performance and other students. Managing collaborative groups may therefore present some initial problems, especially if you have not organized your class in this way before.

Below are some key points to remember as you develop a group approach.

- Explain to students that they are going to work together. Explain *why* ("two heads are better than one" may be a cliché—but it is still relevant).
- Stress the responsibility each group member has to the others in the group.
- Choose student groups carefully to ensure each group has a balance of ability, special talents, gender and ethnicity.
- Make it clear that groups are not permanent and they may change occasionally.
- Help students see the benefits of learning with and from each other.
- Ensure that there are some opportunities for students to work alone (certain activities, writing for example, are more efficiently done in solitude).

Student Discussion

Encourage all students to participate in class discussions. Typically, several students dominate discussion while others hesitate to volunteer comments. Encourage active participation by explicitly stating that you value all students' comments. Reinforce this by not rejecting answers that appear wrong. Ask students to clarify contentious comments. If you ask for students' opinions, be prepared to accept them uncritically.

Assessing Student Learning in Investigating Earth Systems

The completion of the final investigation serves as the primary source of summative assessment information. Traditional assessment strategies often give too much attention to the memorization of terms or the recall of information. As a result, they often fall short of providing information about students' ability to think and reason critically and apply information that they have learned. In *Investigating Earth Systems*, the solutions students provide to the final investigation in each module provide information used to assess thinking, reasoning, and problem-solving skills that are essential to life-long learning.

Assessment is one of the key areas that teachers need to be familiar with and understand when trying to envision implementing *Investigating Earth Systems*. In any curriculum model, the mode of instruction and the mode of assessment are connected. In the best scheme, instruction and assessment are aligned in both content and process. However, to the extent that one becomes an impediment to reform of the other, they can also be uncoupled. *Investigating Earth Systems* uses multiple assessment formats. Some are consistent with reform movements in science education that *Investigating Earth Systems* is designed to promote. **Project-based assessment**, for example, is built into every *Investigating Earth Systems* culminating investigation. At the same time, the developers acknowledge the need to support teachers whose classroom context does not allow them to depart completely from "traditional" assessment formats, such as paper and pencil tests.

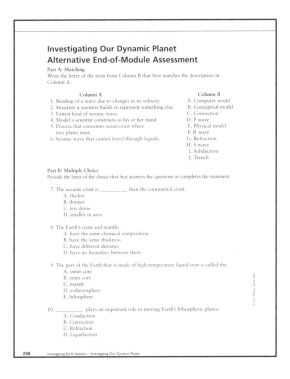

In keeping with the discussion of assessment outlined in the National Science Education Standards (NSES), teachers must be careful while developing the specific expectations for each module. Four issues are of particular importance in that they may present somewhat new considerations for teachers and students. These four issues are dealt with on the next two pages.

1. Integrative Thinking

The National Science Education Standards (NSES) state: "Assessments must be consistent with the decisions they are designed to inform." This means that as a prerequisite to establishing expectations, teachers should consider the use of assessment information. In *Investigating Earth Systems*, students often must be able to articulate the connection between Earth science concepts and their own community. This means that they have to integrate traditional Earth science content with knowledge of their surroundings. It is likely that this kind of integration will be new to students, and that they will require some practice at accomplishing it. Assessment in one module can inform how the next module is approached so that the ability to apply Earth science concepts to local situations is enhanced on an ongoing basis.

2. Importance

An explicit focus of NSES is to promote a shift to deeper instruction on a smaller set of core science concepts and principles. Assessment can support or undermine that intent. It can support it by raising the priority of in-depth treatment of concepts, such as students evaluating the relevance of core concepts to their communities. Assessment can undermine a deep treatment of concepts by encouraging students to parrot back large bodies of knowledge-level facts that are not related to any specific context in particular. In short, by focusing on a few concepts and principles, deemed to be of particularly fundamental importance, assessment can help to overcome a bias toward superficial learning. For example, assessment of terminology that emphasizes deeper understanding of science is that which focuses on the use of terminology as a tool for communicating important ideas. Knowledge of terminology is not an end in itself. Teachers must be watchful that the focus remains on terminology in use, rather than on rote recall of definitions. This is an area that some students will find unusual if their prior science instruction has led them to rely largely on memorization skills for success.

3. Flexibility

Students differ in many ways. Assessment that calls on students to give thoughtful responses must allow for those differences. Some students will find the open-ended character of the *Investigating Earth Systems* module reports disquieting. They may ask many questions to try to find out exactly what the finished product should look like. Teachers will have to give a consistent and repeated message to those students, expressed in many different ways, that the ambiguity inherent in the open-ended character of the assessments is an opportunity for students to show what they know in a way that makes sense to them. This also allows for the assessments to be adapted to students with differing abilities and proficiencies.

4. Consistency

While the module reports are intended to be flexible, they are also intended to be consistent with the manner in which instruction happens, and the kinds of inferences that are going to be made about students' learning on the basis of them. The *Investigating Earth Systems* design is such that students have the opportunity to learn new material in a way that places it in context. Consistent with that, the module reports also call for the new material to be expressed in context. Traditional tests are less likely to allow this kind of expression, and are more likely to be inconsistent with the manner of teaching that *Investigating Earth Systems* is designed to promote. Likewise, in that *Investigating Earth Systems* is meant to help students relate Earth Science to their community, teachers will be using the module reports as the basis for inferences regarding the students' abilities to do that. The design of the module reports is intended to facilitate such inferences.

An assessment instrument can imply but not determine its own best use. This means that *Investigating Earth Systems* teachers can inadvertently assess module reports in ways that work against integrative thinking, a focus on important ideas, flexibility in approach, and consistency between assessment and the inferences made from that assessment.

All expectations should be communicated to students. Discussing the grading criteria and creating a general rubric are critical to student success. Better still, teachers can engage students in modifying and/or creating the criteria that will be used to assess their performance. Start by sharing the sample rubric with students and holding a class discussion. Questions that can be used to focus the discussion include: Why are these criteria included? Which activities will help you to meet these expectations? How much is required? What does an "A" presentation or report look like? The criteria should be revisited throughout the completion of the module, but for now students will have a clearer understanding of the challenge and the expectations they should set for themselves.

Investigating Earth Systems Assessment Tools

Investigating Earth Systems provides you with a variety of tools that you can use to assess student progress in concept development and inquiry skills. The series of evaluation sheets and scoring rubrics provided in the back of this Teacher's Edition should be modified to suit your needs. Once you have settled on the performance levels and criteria and modified them to suit your particular needs, make the evaluation sheets available to students, preferably before they begin their first investigation. Consider photocopying a set of the sheets for each student to include in his or her journal. You can also encourage your students to develop their own rubrics. The final investigation is well-suited for such, since students will have gained valuable experience with criteria by the time they get to this point in the module. Distributing and discussing the evaluation sheets will help students to become familiar with and know the criteria and expectations for their work. If students have a complete set of the evaluation sheets, you can refer to the relevant evaluation sheet at the appropriate point within an *IES* lesson.

1. Pre-Assessment

The pre-assessment activity culminates with students putting their journals together and adding their first journal entry. It is important that this not be graded for content. Credit should be given to all students who make a reasonable attempt to complete the activity. The purpose of this pre-assessment is to provide a benchmark for comparison with later work. At the end of the module, the central questions of the pre-assessment are repeated in the section called **Back to the Beginning**.

2. Assessing the Student Journal

As students complete each investigation, reinforce the need for all observations and data to be organized well and added to the journals. Stress the need for clarity, accurate labeling, dating, and inclusion of all pertinent information. It is important that you assess journals regularly. Students will be more likely to take their journals seriously if you respond to their work. This does not have to be particularly time-consuming. Five types of evaluation instruments for assessing journal entries are available at the back of this Teacher's Edition to help you provide prompt and effective feedback. Each one is explained in turn below.

Journal Entry-Evaluation Sheet

This sheet provides you with general guidelines for assessing student journals. Adapt this sheet so that it is appropriate for your classroom. The journal entry evaluation sheet should be given to students early in the module, discussed with students, and used to provide clear and prompt feedback.

Journal Entry-Checklist

This checklist provides you and your students with a guide for quickly checking the quality and completeness of journal entries. You can assign a value to each criterion, or assign a "+" or "-" for each category, which you can translate into points later. However you choose to do this, the point is to make it easy to respond to students' work quickly and efficiently. Lengthy comments may not be necessary. Depending on time constraints, you may not have time to write comments each time you evaluate journals. The important thing is that students get feedback—they will do better work if they see that you are monitoring their progress.

Key Question Evaluation Sheet

This sheet will help students to learn the basic expectations for the warm-up activity. The **Key Question** is intended to reveal students' conceptions about the phenomena or processes explored in the activity. It is not intended to produce closure, so your assessment of student responses should not be driven by a concern for correctness. Instead, the evaluation sheet emphasizes that you want to see evidence of prior knowledge and that students should communicate their thinking clearly. It is unlikely that you will have time to apply this assessment every time students complete a warm-up activity, yet in order to ensure that students value committing their initial conceptions to paper and taking the warm-up seriously, you should always remind students of the criteria. When time permits, use this evaluation sheet as a spot check on the quality of their work.

Investigation Journal Entry-Evaluation Sheet

This sheet will help students to learn the basic expectations for journal entries that feature the write-up of investigations. *IES* investigations are intended to help students to develop content understanding and inquiry abilities. This evaluation sheet provides a variety of criteria that students can use to ensure that their work meets the highest possible standards and expectations. When assessing student investigations, keep in mind that the **Investigate** section of an *IES* lesson corresponds to the explore phase of the learning cycle (engage, explore, apply, evaluate) in which students explore their conceptions of phenomena through hands-on activity. Using and discussing the evaluation sheet will help your students to internalize the criteria for their performance. You can further encourage students to internalize the criteria by making the criteria part of your "assessment conversations" with them as you circulate around the classroom and discuss student work. For example, while students are working, you can ask them criteria-driven questions such as: "Is your work thorough and complete? Are all of you participating in the activity? Do you each have a role to play in solving the problem?" and so on.

Review and Reflect Journal Entry-Evaluation Sheet

Reviewing and reflecting upon one's work is an important part of scientific inquiry and is also important to learning science. Depending upon whether you have students complete the work individually or within a group, the **Review and Reflect** portion of each investigation can be used to provide information about individual or collective understandings about the concepts and inquiry processes explored in the investigation. Whatever choice you make, this evaluation sheet provides you with a few general criteria for assessing content and thoroughness of student work. Adapt and modify the sheet to meet your needs. Consider involving students in selecting and modifying the criteria for evaluating their end of investigation reflections.

3. Assessing Group Participation

One of the challenges to assessing students who work in collaborative teams is assessing group participation. Students need to know that each group member must pull his or her weight. As a component of a complete assessment system, especially in a collaborative learning environment, it is often helpful to engage students in a self-assessment of their participation in a group. Knowing that their contributions to the group will be evaluated provides an additional motivational tool to keep students constructively engaged. These evaluation forms (Group Participation Evaluation Sheets I and II) provide students with an opportunity to assess group participation. In no case should the results of this evaluation be used as the sole source of assessment data. Rather, it is better to assign a weight to the results of this evaluation and factor it in with other sources of assessment data. If you have not done this before, you may be surprised to find how honestly students will critique their own work, often more intensely than you might do.

4. Assessing the Final Investigation

Students' work throughout the module culminates with the final investigation. To complete it, students need a working knowledge of previous activities. Because it refers back to the previous steps, the last investigation is a good review and a chance to demonstrate proficiency. For an idea on how to use the last investigation as a performance-based exam, see the section in the **Appendix**.

5. Assessing Inquiry Processes

There is an obvious difficulty in assessing individual student proficiency when the students work within a collaborative group. One way to do this is to have a group present its results followed by a question-and-answer session. You can direct questions to individual students as a way of checking proficiency. Another is to have every student write a report on his or her role in the investigation, after first making it clear what this report should contain. Individual interviews are clearly the best option but may not be feasible given the time constraints of most classes.

6. Traditional Assessment Options

A traditional paper-and pencil-exam is included in the **Appendices**. While performance-based assessments may offer teachers more insight into student skill levels, computer-generated tests are also useful—especially so in states with state-sponsored exams. Additionally, some students are strong in one area and not as strong in another. Using a variety of methods for assessing and grading students' progress offers a more complete picture of the success of the student—and the teacher.

Reviewing and Reflecting upon Your Teaching

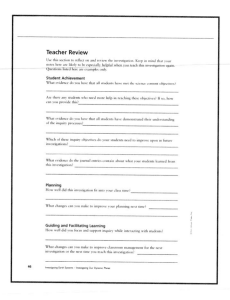
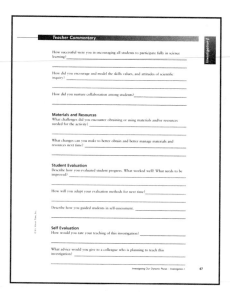

Reviewing and Reflecting upon Your Teaching provides an important opportunity for professional growth. A two-page Teacher Review form is included at the end of each investigation. The purpose of these reviews is to help you to reflect on your teaching of each investigation. We suggest that you try to answer each question at the completion of each investigation, then go back to the relevant section of this Teacher's Edition and write specific comments in the margins. Use the comments the next time you teach the investigation. For example, if you found that you were able to make substitutions to the list of materials needed, write a note about those changes in the margin of that page of this Teacher's Edition.

 GETIT™ Geoscience Education Through Interactive Technology for Grades 6-12

Earthquakes, volcanoes, hurricanes, and plate tectonics are all subjects that deal with energy transfer at or below the Earth's surface. The GETIT CD-ROM uses these events to teach the fundamentals of the Earth's dynamism. GETIT contains 63 interactive—inquiry-based—activities that closely simulate real-life science practice. Students work with real data and are encouraged to make their own discoveries—often learning from their mistakes. They use an electronic notebook to answer questions and record ideas, and teachers can monitor their progress using the integrated class-management module. The Teacher's Guide includes Assessments, Evaluation Criteria, Scientific Content, Graphs, Diagrams and Blackline Masters. GETIT conforms to the National Science Education Standards and the American Association for the Advancement of Science benchmarks for Earth Science.

Enhancing *Investigating Our Dynamic Planet* with **GETIT**

Investigating Our Dynamic Planet				
Investigation	Key Question	Page	Investigate or question	GETIT Activity
1. Gathering Evidence and Modeling	How do you make a model of something that you cannot see?	P6	Numerical models	• Does pressure influence volume? • Does volume influence density? • Does temperature influence volume?
2. The Interior of the Earth	What is the interior of the Earth like?	P11	Part B: Kinds of seismic waves	• Seismic wave properties
		P12	Part C: Refraction of waves	• Earth's core casts a big shadow
		P15	Part D: Refraction of earthquake waves in the Earth	• Earth's core casts a big shadow • Journey through the center of Earth
		P16	What Earthquake waves reveal about the interior of the Earth	• Seismic wave properties
		P17	Earthquakes and seismic waves	• How much energy is released by an earthquake? • Seismic wave properties • How much damage is done? • Science Showtime episode: Whole lotta shakin' going on

Enhancing *Investigating Our Dynamic Planet* with **GETIT**

Investigating Our Dynamic Planet				
Investigation	Key Question	Page	Investigate or question	GETIT Activity
3. Forces that Cause Earth Movements	Does the rock of the Earth's mantle move?	P25	**Digging Deeper:** Convection	• Where's the heat? • Density, the basics • Does temperature influence volume? • Does volume influence density? • What affects rock melting? • Science Showtime episode: The pressure is off
		P27	**Digging Deeper:** Mid-Ocean ridges	• The pressure is off
4. The Movement of the Earth's Lithospheric Plates	What happens where lithospheric plates meet?	P30	**Part A:** Modeling plate convergence	• Science Showtime episode: Whole lotta shakin' going on • Plate reconstruction
		P34	**Part B:** Modeling plate boundaries	• A tale of three margins • Earthquakes, volcanoes, and plate margins
		P35	**Digging Deeper**	• Science Showtime episode: You're in hot water now
5. Earthquakes, Volcanoes, and Mountains	How are earthquakes, volcanoes, and mountains related?	P41	**Investigate** and **Digging Deeper**	• Quick earthquake locator • Significant earthquakes • Global earthquake locator • Global volcano locator • Plate tectonics • Where's the heat? • What is volcanic activity? • Hawaii • Types of volcanoes • Eruptive styles
6. Earth's Moving Continents	Have the continents and oceans always been in the position they are today?	P52	Investigate	• Plate reconstruction
7. Natural Hazards and Our Dynamic Planet	What natural hazards do dynamic events cause?	P62	Investigate	• What is volcanic activity? • Are all volcanic eruptions equal?
		P65	**Digging Deeper**	• How much damage is done?
		P67	Volcano hazards	• Types of volcanoes • Eruptive styles
		P68	Predicting earthquakes and volcanoes.	• Can we use a volcano's past to predict its future?

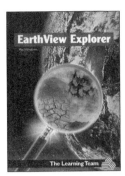

Correlations between IES *Our Dynamic Planet (ODP)*
and *EarthView Explorer (EView)* software
Raymond Sambrotto; Lamont-Doherty Earth Observatory of Columbia University;
Palisades, NY 10964. 845-365-8402.

The content of the *EView* Geosphere unit corresponds well with IES *Our Dynamic Planet*. Both are organized around the central process of sea-floor spreading and plate tectonics and the geological features they produce. In many cases, the similarity extends to the diagrams and maps used in the two titles. Thus, you can easily use *EView* to augment your teaching of the key Earth science concepts explored in *Our Dynamic Planet*.

Correlations between *Our Dynamic Planet* with *EarthView Explorer Software*

IES Our Dynamic Planet	Page	EarthView	Usage notes
Investigation 1: Gathering Evidence and Modeling	P1	Climate unit – Geosphere	*EView's* numerical model of atmospheric temperature in the Climate unit will help students to further explore the concept of modeling Earth processes. *EView's* temperature model can be compared with the mystery bag activity used in the *ODP* **Investigation** and addressed in the **Review and Reflect** section.
Investigation 2: The Interior of the Earth	P8	Geosphere Unit – Earthquake patterns Sea-floor spreading	*EView* covers the distribution of earthquakes but does not cover seismic waves or their propagation. In the **Information** section of *EView's* Earthquake patterns activity, the use of a seismometer is discussed. This links to *ODP's* detailed treatment of wave phenomena, particularly in *ODP's* **Investigate Section A** and **B**. In the **Information** section of *EView's* Sea-floor spreading activity, the first figure shows how seismic waves propagate through the Earth and this links to the **Digging Deeper** reading section of the *ODP* investigation.
Investigation 3: Forces that Cause Earth Movements	P22	Geosphere Unit – Sea-floor spreading	The figures of lithosphere – mantle interaction are almost identical in *EView* and *ODP*. The content in *EView's* Information sections extends the hands-on activity on convection in *ODP's* **Investigate** section. The extensive data sets in *EView* can also be used to illustrate and extend the ideas discussed in *ODP's* **Digging Deeper** reading section.
Investigation 4: The Movement of the Earth's Lithospheric Plates	P30	Geosphere Unit – Earthquake patterns Volcanoes Sea-floor spreading Climate Unit – Geosphere	*EView's* Earthquake patterns and Volcanoes activities provide extensive data to illustrate the location and characteristics of earthquakes explored in *ODP's* Investigate section. Like *ODP's* **Digging Deeper** reading section, *EView's* sea-floor spreading activity links the phenomena of earthquakes, volcanoes, and mountains with the underlying process of plate tectonics.
Investigation 5: Earthquakes, Volcanoes, and Mountains	P41	Geosphere Unit – Earthquake patterns Volcanoes Sea-floor spreading	*EView's* Earthquake patterns and Volcanoes activities provide extensive data to illustrate the location and characteristics of earthquakes. These can be used to supplement student exploration in ODP's Investigate section. Like *ODP's* **Digging Deeper** reading section, *EView's* sea-floor spreading activity links the phenomena of earthquakes, volcanoes, and mountains with the underlying process of plate tectonics.

Correlations between *Our Dynamic Planet* with *EarthView Explorer Software*

IES Our Dynamic Planet	Page	EarthView	Usage notes
Investigation 6: Earth's Moving Continents	P51	Geosphere Unit – Sea-floor spreading	Like *ODP*, *EView* explores the evidence for plate tectonics and continental drift. Data explorations in *EView* can be used in concert with illustrations and verbal descriptions provided in *ODP*.
Investigation 7: Natural Hazards and Our Dynamic Planet	P61	Geosphere Unit – Earthquake patterns Volcanoes	This *ODP* **Investigation** focuses on the societal relevance of lithospheric processes that are also addressed in the two *EView* activities listed. In particular, the *EView* activities contain population data that can be used to assess hazards and risk that are addressed in *ODP's* **Investigate** section. Also, *EView* activities support the concept of hazards developed in the *ODP* **Digging Deeper** reading section.

Investigating Our Dynamic Planet: Introduction

From day to day, and from year to year, we see little change in the rocks and the terrain of the Earth, yet we know that the rocks exposed at the Earth's surface are gradually being weathered into particles, large and small, and dissolved materials. These products reside for a time as soils on the land surface, but they are ultimately carried by streams and rivers, and in some places by wind or by moving glacier ice, to resting places on the ocean bottom. Over times of hundreds of thousands to millions of years, these slow processes result in great changes in the Earth's surface environment. Geologic time—measured in billions of years— is unimaginably longer than even these seemingly long time scales. This points out how dynamic the Earth is, even without consideration of the occasional violent events like earthquakes and volcanoes, which have durations that are short even by human standards.

As a consequence of the first great revolution in geology, brought about by the Scottish natural scientist James Hutton, geologists became convinced that the history of the Earth, recorded in the rocks of the continents, could be interpreted in terms of events and processes of the kind we can observe and study today. Until the middle of the twentieth century, however, with the advent of the theory of plate tectonics, geologists lacked the key to understanding the driving mechanisms in the Earth's interior that can account for the processes we see acting to shape the Earth.

After the second great revolution in geology, that of the theory of plate tectonics, geoscientists had the key to understanding the dynamics of the Earth's interior. Mantle convection, caused by heating from below, by the hotter core, and cooling from above, by the cool surface of the Earth, drives the motions of the Earth's lithospheric plates. Downward to depths of tens to as much as 200 kilometers, the rocks of the Earth are cool enough to remain rigid, whereas deeper in the mantle, the rocks flow plastically as part of large-scale convection cells. Speeds of movement range from a few centimeters per year to as much as 20 centimeters per year. New lithospheric-plate material is produced at the mid-ocean ridges, where partial melting of ascending, hot mantle rock feeds submarine volcanoes at the ridge crests. The newly created plates then move toward subduction zones, where one plate dives down another plate into the deep mantle, ultimately to be resorbed into the mantle.

Where the rate of subduction becomes greater than the rate of plate production at the ridge, the ridge is eventually swallowed down the subduction zone, never to be seen again. Thereafter, the ocean narrows, and eventually closes up, resulting in a continent–continent collision at the site of the subduction zone. These processes of subduction and collision have shaped much of the geologic record of the continents.

Subduction zones are the sites of frequent earthquakes and continual volcanic activity. Major earthquakes are generated as the downgoing rigid plate is subjected to various internal forces. These earthquakes happen to depths of as much as several hundred kilometers. Wide areas lying adjacent to the subduction zone and above the downgoing plate are vulnerable to repeated major earthquakes. As the plate descends to great depths and is gradually heated, much of the water that was originally incorporated into the rocks of the plate is released, and as it rises upward, it causes partial melting of the mantle rock above the subduction zone. In this way, long chains of active volcanoes are formed at a certain distance away from the locus of subduction. The rim of the Pacific Ocean is in most places the site of subduction zones, as the other oceans of the world expand at the expense of the Pacific. Most of the world's great earthquakes and volcanic eruptions happen around the Pacific rim.

More Information...on the Web

Go to the *Investigating Earth Systems* web site www.agiweb.org/ies for links to a variety of other web sites that will help you deepen your understanding of content and prepare you to teach this module, *Investigating Our Dynamic Planet*.

Students' Conceptions about Our Dynamic Planet

While most students will have some clear ideas about what the surface of the Earth is like, they are unlikely to know that Earth's crust and outer mantle act as a system that scientists describe as the geosphere. Some students may have heard of the term plate tectonics and connect it with earthquakes and volcanoes, but these are likely to be only informal ideas and not linked to a conceptual understanding of Earth's interior operating as a dynamic system.

Students commonly associate the occurrence of earthquakes and volcanoes with pressure. Some students believe that earthquakes cause volcanoes and/or that volcanoes cause earthquakes. Students often associate earthquakes with faults, but their ideas about the nature of the relationship between a fault and earthquake are often undeveloped. For example, they may claim that faults cause earthquakes or that earthquakes cause faults and move directly into a description of the destruction caused by the earthquakes without explaining why the earthquake caused the fault or vice versa. Middle-school students rarely describe faults as a surface, but some students note that forces at work on opposite sides of the fault cause pressure, which leads to earthquakes. Students typically associate volcanoes with pressure and/or heat, but it is often difficult to sort out the exact nature of their understanding. Middle-school students commonly state that heat can cause pressure, or that pressure can cause heat, and that either or both make magma move upward toward the surface. Students' conceptions about how mountains form include erosion, the buildup of layers of rock over time, volcanic eruption (very common), and plates moving toward one another and forcing the land upward.

Field testing of the program revealed that middle-school students have a variety of ideas about Earth dynamics. Not all students have the same ideas, but the following sampling of responses will give you a general sense of what you might expect middle-school students to know about the causes of volcanoes, earthquakes, and mountains as they come into the module. A more extensive sampling of responses is presented in the pre-assessment section of this Teacher's Edition.

Ideas about what volcanoes are and why they occur where they do:
Heat comes from the outer core and comes up and splits the Earth. Molten lava comes to the volcano and then pressure pushes it out.
Volcanoes are mountains that stay active because they have a magma chamber under them, which pushes heat and lava up and creates deadly gases in the air.
The heat from the mantle makes volcanoes erupt. The heat and lava get pushed up and it can't hold it anymore so it erupts and causes mountains to form.
Underneath the ground the magma gets so hot it has nowhere to go but up, so then the volcano erupts.

Ideas about what earthquakes are and what causes them:
A fault in the ground causes the ground to start shaking.
Below the ground is a fault. I am not sure how it happens, but above ground houses fall apart, trees break, and buildings collapse.
The friction between two plates moving in the opposite way makes an earthquake.

Ideas about how earthquakes and volcanoes are related:
Earthquakes and volcanoes happen near each other most often and most are on the Pacific Ocean.
Volcanoes make earthquakes, but earthquakes make volcanoes erupt.
Volcanoes and earthquakes sometimes occur at the same time.

Ideas about how mountains form:
The wind wore the land into mountains after the ancient ocean covered them.
A mountain is formed by molten rock after a volcano has occurred.
Mountains are caused by the land shifting and pushing together, causing the land to go up.

It is crucial that you find out what informal ideas your students already have about *Our Dynamic Planet* before beginning this module. The pre-assessment activity will tell you much of what you need to know in addressing your students' unique needs.

Investigating Our Dynamic Planet: Module Flow

Activity Summaries	Emphasis
Pre-Assessment Students describe their understanding of key concepts explored in the module.	Recording initial knowledge and understanding of content.
Introducing *Our Dynamic Planet* Students discuss their ideas and experiences related to the topics they will be investigating.	Putting the investigations into a meaningful context.
Investigation 1: Gathering Evidence and Modeling Students investigate the idea of modeling something that cannot be seen, using their senses to make observations about a bag's hidden contents.	Modeling, using senses, making observations and inferences. Raising questions through simulations, constructing and using models, recording observations, and sharing findings.
Investigation 2: The Interior of the Earth Students investigate wave speed (using water), kinds of waves (using Slinkys®), and wave refraction to explore how scientists have developed models of the Earth's interior structure.	Using tools, applying mathematics, devising and carrying out a plan, and sharing findings.
Investigation 3: Forces that Cause Earth Movements Students investigate convection in a simple hands-on activity and relate their model to the process of mantle convection, deepening their understanding of the Earth's dynamic interior.	Investigating natural processes using models, and making diagrams.
Investigation 4: The Movement of the Earth's Lithospheric Plates Students begin to explore the surface processes and features that result from plate motion by using use simple materials to simulate motions and events at plate boundaries.	Investigating natural processes using models, interpreting findings, and sharing findings.
Investigation 5: Earthquakes, Volcanoes, and Mountains Students plot the locations of earthquakes, volcanoes, and mountain ranges on a world map to explore the relationships between these event and features, and plate tectonics.	Synthesizing data onto maps, using maps and data tables as scientific tools, searching for patterns and relationships, conducting research, and sharing findings.
Investigation 6: Earth's Moving Continents Students explore several lines of geologic and geographical evidence for continental drift. Reviewing the 250-million-year history of the breakup of Pangea helps students to understand that our planet has a long history of dynamic change through time.	Exploring evidence, interpreting evidence, and sharing findings.
Investigation 7: Natural Hazards and Our Dynamic Planet Students use their knowledge and skills about our dynamic planet to investigate the hazards that result from plate motion and history of dynamic change through time.	Synthesizing results, conducting research, applying results of previous experiments, and communicating findings.
Reflecting Students review the science content and inquiry processes they used throughout the module.	Assessing student learning.

Investigating Our Dynamic Planet: Module Objectives

Investigation	Science Content	Inquiry Process Skills
Investigation 1: Gathering Evidence and Modeling Students make and revise models of the contents within a "mystery bag".	Students will collect evidence that: 1. Something that cannot be seen can be modeled. 2. Scientists use the words "model" and "modeling" in different ways. 3. Models can be physical, conceptual, mathematical, or numerical. 4. Models are often revised as new evidence is gathered.	Students will: 1. Make observations using the senses. 2. Record observations in a systematic way. 3. Make a model of what they cannot see. 4. Revise models on the basis of additional observations. 5. Communicate observations and findings to others.
Investigation 2: The Interior of the Earth Students investigate wave speed (using water), kinds of waves (using Slinkys®), and wave refraction to become familiar with the evidence that scientists have used to model the Earth's interior.	Students will collect evidence that: 1. Waves travel at a certain rate in a uniform material. 2. Waves carry energy with them as they move through a material. 3. Waves are refracted when they enter material of a different density. 4. Compressional waves can travel through solids, liquids, and gases, but shear waves can travel only through solids. 5. The Earth's interior has a layered structure including crust, mantle, and core, which we have inferred from the study of seismic waves. 6. Models can be revised and improved with additional data.	Students will: 1. Predict the behavior of waves in water. 2. Conduct fair and objective tests. 3. Measure distance and time. 4. Calculate rate. 5. Make observations using the senses. 6. Record observations. 7. Devise a plan to investigate a question. 8. Collect observational data. 9. Compare data sets. 10. Use test data to interpret a model. 11. Communicate observations and findings to others.
Investigation 3: Forces that Cause Earth Movements Students investigate convection in a simple hands-on activity and relate their model to the process of mantle convection.	Students will collect evidence that: 1. Convection is a motion in a fluid that results from the fluid being heated from below and cooled from above. 2. Earth has an outer rigid shell (the lithosphere) underlain by a layer of material (the asthenosphere) that is hot and under enough pressure to deform slowly over time. 3. Convection currents, driven by uneven heating within the Earth, are associated with movement of lithospheric plates. 4. New crust is formed where plates spread apart.	Students will: 1. Use models to investigate questions about plate movement. 2. Compare aspects of the model with the actual event or phenomenon. 3. Apply knowledge from modeling and model analysis to a new situation. 4. Arrive at conclusions based on data analysis. 5. Communicate findings and results to others.

Investigating Our Dynamic Planet: Module Objectives

Investigation	Science Content	Inquiry Process Skills
Investigation 4: The Movement of the Earth's Lithospheric Plates Students begin to explore the surface processes and features that result from plate motion by using use simple materials to simulate motions and events at plate boundaries.	Students will collect evidence that: 1. The Earth's crust consists of thick, less dense continental crust and thin, more dense oceanic crust. 2. The lithosphere is not one continuous piece, but instead exists as large and small pieces or plates. 3. Plates can be moving apart from one another, moving toward one another, or sliding past one another. 4. Plates with ocean crust are denser and slide under plates with continental crust when they collide. 5. Earthquakes, mountains, and/or volcanoes often occur at the boundaries between plates.	Students will: 1. Use models to investigate a science question. 2. Make predictions about the outcome of the modeling events. 3. Collect data from the models. 4. Compare data from the models. 5. Use knowledge from a model data analysis to interpret maps. 6. Share findings with others.
Investigation 5: Earthquakes, Volcanoes, and Mountains Students plot the locations of earthquakes, volcanoes, and mountain ranges on a world map to explore the relationships between these events and features, and plate tectonics.	Students will collect evidence that: 1. Earthquakes occur when rocks on either side of a fault slide past one another. 2. Volcanoes result from the eruption of molten rock, volcanic fragments, and gases at the Earth's surface. 3. Gas content often controls the explosiveness of a volcanic eruption. 4. Most earthquakes and volcanoes occur along plate boundaries. 5. Most of the world's major mountain chains are formed where two lithospheric plates converge.	Students will: 1. Interpret data sets. 2. Plot coordinates on maps. 3. Look for patterns and relationships in data sets. 4. Draw conclusions about relationships between data sets. 5. Conduct additional research on earthquakes, volcanoes, and mountain chains. 6. Collate information into a useful format. 7. Communicate observations and findings to others.
Investigation 6: Earth's Moving Continents Students explore several lines of geologic and geographical evidence for continental drift during the last 250 million years.	Students will collect evidence that: 1. Continent shapes appear to fit together. 2. There are common fossils, mountain chains, and glacial deposits on different continents, some of which are now widely separated by oceans. 3. The theory of continental drift fell out of favor with scientists until a mechanism (plate tectonics) for moving the continents was discovered.	Students will: 1. Collect evidence from maps and other sources of information to support or refute a theory. 2. Analyze data, looking for patterns and relationships. 3. Use data analyses to support or refute a theory. 4. Revise conclusions on the basis of new evidence. 5. Share findings and conclusions with others.

Investigating Our Dynamic Planet: Module Objectives

Investigation	Science Content	Inquiry Process Skills
Investigation 7: Natural Hazards and Our Dynamic Planet Students use their knowledge and skills about our dynamic planet to create an informative brochure for a community located near an earthquake or volcano hazard site.	Students will collect evidence that: 1. Natural hazards have an impact on communities. 2. Natural hazards like earthquakes and volcanoes are more likely to happen in some areas than in others. 3. If well informed, humans can prepare themselves for natural disasters.	Students will: 1. Use data and information about natural disasters to create a brochure. 2. Analyze which data are suitable for inclusion in the brochure. 3. Conduct additional research on natural disasters. 4. Collate information into a useful format. 5. Communicate to others what they have discovered about natural disasters and how to deal with them.

National Science Education Content Standards

Investigating Earth Systems is a Standards-driven curriculum. That is, the scope and sequence of the series is derived from, and driven by, the National Science Education Standards (NSES) and the American Association for the Advancement of Science (AAAS) Benchmarks for Science Literacy (BSL). Both specify content standards that students should know by the completion of eighth grade.

Unifying Concepts and Processes
- Systems, order, and organization
- Evidence, models, and explanation
- Constancy, change, and measurement
- Evolution and equilibrium

Science as Inquiry
- Identify questions that can be answered through scientific investigations
- Design and conduct a scientific investigation
- Use tools and techniques to gather, analyze, and interpret data
- Develop descriptions, explanations, predictions, and models based upon evidence
- Think critically and logically to make the relationships between evidence and explanation
- Recognize and analyze alternative explanations and predictions
- Communicate scientific procedures and explanations
- Use mathematics in all aspects of scientific inquiry
- Understand scientific inquiry

Physical Science
- Properties and changes of properties in matter
- Motion and forces
- Transfer of energy

Life Science
- Diversity and adaptation of organisms

Earth and Space Science
- Structure of the Earth system
- Earth's history
- Earth in the Solar System

Science and Technology
- Abilities of technological design
- Understandings about science and technology

Science in Personal and Social Perspectives
- Natural hazards
- Risks and benefits
- Science and technology in society

History and Nature of Science
- Science as a human endeavor
- Nature of science
- History of science

Key NSES Earth Science Standards Addressed in IES Our Dynamic Planet

1. The solid Earth is layered with the lithosphere; hot, convecting mantle; and dense, metallic core.
2. Lithospheric plates on the scales of continents and oceans consistently move at the rate of centimeters per year in response to movements in the mantle. Major geological events, such as earthquakes, volcanic eruptions, and mountain building, result from these plate motions.
3. Landforms are the result of a combination of constructive and destructive forces. Constructive forces include crustal deformation, volcanic eruptions, and deposition of sediment, while destructive forces include weathering and erosion.
4. Some changes in the solid Earth can be described as the "rock cycle." Old rocks at the Earth's surface weather, forming sediments that are buried, then compacted, heated, and often recrystallized into new rock. Eventually, those new rocks may be brought to the surface by the forces that drive plate motions, and the rock cycle continues.
5. The Earth processes we see today, including erosion, movement of the lithospheric plates, and changes in atmospheric composition, are similar to those that occurred in the past.
6. Fossils provide important evidence of how life and environmental conditions have changed.
7. Natural hazards include earthquakes, landslides, wildfires, volcanic eruptions, floods, storms, and even possible impacts of asteroids.
8. Students should understand the risks associated with natural hazards (fires, flood, tornadoes, hurricanes, earthquakes, and volcanic eruptions), with chemical hazards (pollutants in air, water, soil and food).

Other NSES Content Standards Addressed in IES Our Dynamic Planet

1. Evidence consists of observations and data on which to base scientific explanations.
2. Models are tentative schemes or structures that correspond to real objects, events, or classes of events, and that have explanatory power. Models help scientists and engineers understand how things work. Models take many forms, including physical objects, plans, mental constructs, mathematical equations, and computer simulations.

Key AAAS Earth Science Benchmarks Addressed in IES Our Dynamic Planet

The Physical Setting, Section B: The Earth

1. The Earth is mostly rock. Three-fourths of its surface is covered by a relatively thin layer of water (some of it frozen), and the entire planet is surrounded by a relatively thin blanket of air. It is the only body in the solar system that appears able to support life. The other planets have compositions and conditions very different from the Earth's.

The Physical Setting, Section B: Processes that Shape the Earth

1. The interior of the Earth is hot. Heat flow and movement of material within the Earth cause earthquakes and volcanic eruptions and create mountains and ocean basins. Gas and dust from large volcanoes can change the atmosphere.

2. Some changes in the Earth's surface are abrupt (such as earthquakes and volcanic eruptions) while other changes happen very slowly (such as uplift and wearing down of mountains). The Earth's surface is shaped in part by the motion of water and wind over very long times, which act to level mountain ranges.

3. Thousands of layers of sedimentary rock confirm the long history of the changing surface of the Earth and the changing life forms whose remains are found in successive layers. The youngest layers are not always found on top, because of folding, breaking, and uplift of layers.

The Physical Setting, Section F: Motion

1. Vibrations in materials set up wavelike disturbances that spread away from the source. Sound and earthquake waves are examples. These and other waves move at different speeds in different materials.

Common Themes, Section B: Models

1. Models are often used to think about processes that happen too slowly, too quickly, or on too small a scale to be observed directly, or that are too vast to be changed deliberately, or that are potentially dangerous.

Materials and Equipment List for Investigating Our Dynamic Planet

Pre-Assessment

Each group of students will need:

- poster board, poster paper, or butcher paper
- Student Journal cover sheet, one for each student (**Blackline Master** *Our Dynamic Planet* P.2, available at the back of this Teacher's Edition)

Teachers will need:

- overhead projector, chalkboard, or flip-chart paper
- transparency of **Blackline Master** *Our Dynamic Planet* P.1 (Questions about Our Dynamic Planet)

Materials Needed for Each Group per Investigation

Investigation 1

- double-bagged, brown-paper bag with "mystery objects" sealed inside
- colored pencils

Investigation 2

Part A

- large, flat-bottomed container   ← *pyrex (from home)*
- black permanent marker
- plastic metric ruler – 30 cm
- pebble about one centimeter in diameter
- flashlight
- stopwatch

Part B

- two Slinkys®

Part C

- large open floor area (paved parking lot or playground)
- piece of white chalk (or masking tape)
- piece of red chalk
- clear container
- long pencil
- Photocopy of **Blackline Master** *Our Dynamic Planet* 2.2 Modeling Wave Refraction

Part D (material needed per person)

- pencil with a good eraser
- transparent straightedge or ruler
- Photocopy of **Blackline Master** *Our Dynamic Planet* 2.3 Refraction of Earthquake Waves

Investigation 3

- candle
- small heat-resistant container (empty, clean tuna fish can, label removed)
- corn syrup (or pancake syrup)
- two pieces of thin cardboard, 1 cm square
- two small bricks
- lighter or matches

For the demonstration (**Step 5** of the **Investigation**), the teacher will need the following materials:

- hot plate
- large clear heatproof beaker (1L)
- cup of oatmeal
- food coloring
- sawdust

Investigation 4

Part A

- scissors
- thick corrugated cardboard
- thin cardboard (like a cereal box)
- duct tape
- foaming shaving cream
- metric ruler

Part B
Materials to be selected by teacher and students

Investigation 5

- colored pencils (3 different colors)
- copy of **Blackline Master** *Our Dynamic Planet* 5.1 – World Map

Investigation 6

- a copy of **Blackline Master** *Our Dynamic Planet* 6.1 – World Map of Continents and Continental Shelves
- scissors
- construction paper
- glue

Investigation 7

- reference materials *
- access to computer word processing and desktop publishing (if possible)
- range of general craft materials required to make a brochure

General Supplies

Although the investigations can be done with the specific materials listed, it is always a good idea to build up a supply of general materials.

- 2 or 3 large clear plastic storage bins about 30 cm x 45 cm x 30 cm deep, with lids (these can be used for storage and also make good water containers)
- 2 or 3 plastic buckets and one large water container (camping type with a faucet) ⟵
- rolls of masking tape, duct tape, and clear adhesive tape
- rolls of plastic wrap and aluminum foil
- clear self-locking plastic bags (various sizes)
- ball of string and spools of sewing thread
- pieces of wire (can be pieces of wire coat hangers)
- stapler, staples, paper clips, and binder fasteners
- safety scissors and one sharp knife
- cotton balls, tongue depressors
- plastic and paper cups of various sizes
- empty coffee and soup cans, empty boxes and egg cartons
- several clear plastic soda bottles (various sizes)
- poster board, overhead transparencies, tracing paper, and graph paper
- balances and/or scales, weights, spring scales
- graduated cylinders, hot plates, microscopes
- safety goggles
- disposable latex gloves
- lab aprons or old shirts
- first aid kit

* The *Investigating Earth Systems* web site www.agiweb.org/ies provides topical Internet sites and a list of resources that will aid student research.

Pre-assessment

Overview

During the pre-assessment phase, the students complete an open-ended survey of their knowledge and understanding of key concepts explored in the *Our Dynamic Planet* module. Students are given four questions to consider and their responses become the first entry in their journal.

Preparation and Materials Needed

This pre-assessment activity does not appear in the Student Book. However, to find out what your students already know about our dynamic planet, it is crucial that you conduct this pre-assessment before introducing the module and distributing the Student Books. Make sure that your students understand clearly that the pre-assessment is not a test. Explain to them that, by reviewing the pre-assessment at the end of the module, they can compare how their ideas and knowledge about our dynamic planet have changed as a result of their investigations. Tell them that the pre-assessment will also help you gauge how successful the investigations have been for everyone.

After the pre-assessment but before distributing the Student Books, take some time to reflect on the ideas your students have. This is the starting point. You need to ensure that what follows fits with your students' prior knowledge.

Materials
- poster board, poster paper, or butcher paper
- overhead projector, blackboard, or flip-chart paper
- overhead transparency of questions (**Blackline Master** *Our Dynamic Planet* P.1)
- student journal cover sheet, one for each student (**Blackline Master** *Our Dynamic Planet* P.2)

Suggested Teaching Procedure

1. Let students know that what they write in this exercise will become their first entry in a scientific journal that they will keep throughout the module. Explain that each person is going to write down all the ideas that they have about the answers to questions dealing with our dynamic planet. The reason for this is to provide them, and you, with a starting point for their investigations. Tell students that when they have finished the module, they will answer these same questions again. This will allow them, and you, to compare how their knowledge about our dynamic planet has changed as a result of their investigations.

2. Display the pre-assessment questions on an overhead projector, or provide each student with a copy of the questions (**Blackline Master** *Our Dynamic Planet* P.1). Have students write responses to these questions in their journals.

pg 310

> - What are volcanoes and why do they occur where they do?
> - What are earthquakes and what causes them?
> - How are earthquakes and volcanoes related?
> - How do mountains form?

How do you know it enough detail? well stated, specific, distinctive

Allow a reasonable period of time for all students to respond. Circulate around the classroom, prompting students to provide as much detail as possible.

Sample Student Responses

Ideas about what volcanoes are and why they occur where they do:

How would you describe? Look like? Feel like? Smell like

- The Earth has a lot of pressure and needs to release it.
- Heat comes from the outer core and comes up and splits the Earth. Molten lava comes to the volcano and then pressure pushes it out.
- The mantle turns into molten lava and the pressure causes it to erupt.
- Volcanoes are mountains that stay active because they have a magma chamber under them which pushes heat and lava up and creates deadly gases in the air.
- The heat from the mantle makes volcanoes erupt. The heat and lava get pushed up and it can't hold it anymore so it erupts and causes mountains to form.
- Something under the ground builds up the lava and it gets so full and moves too quickly so it blows up. Ashes, smoke, fire, and lava all come out of the volcano.
- A volcano lets pressure out of the Earth and bends up rock layers and soil. *gases vapor*
- Underneath the ground the magma gets so hot it has nowhere to go but up, so then the volcano erupts. *why?*

Ideas about what earthquakes are and what causes them:

ll

- A fault in the ground causes the ground to start shaking.
- When a fault moves, the earthquake happens. Everything shakes.
- Below the ground is a fault. I am not sure how it happens, but above ground houses fall apart, trees break, and buildings collapse.
- The ground has a lot of pressure. It needs to let it out so it does and it destroys towns.
- As the plates shift, the earthquake starts, creating a fault.
- Two or more plates shift together or apart from one another to form an earthquake.
- The friction between two plates moving in the opposite way makes an earthquake.
- Faults bump together causing the ground above and below to shake.

Ideas about how earthquakes and volcanoes are related:

- When a volcano erupts, the force of the lava exploding may cause an earthquake.
- Earthquakes and volcanoes happen near each other most often and most are on the Pacific Ocean.
- An earthquake forms a volcano.
- Volcanoes and earthquakes can happen on Earth's plates.
- Volcanoes make earthquakes, but earthquakes make volcanoes erupt.
- Volcanoes and earthquakes sometimes occur at the same time.
- Volcanoes and earthquakes are both caused by pressure.
- Earthquakes and volcanoes both cause destruction.

Ideas about how mountains form:

- The wind wore the land into mountains after the ancient ocean covered them.
- The eruption happens, then the molten lava hardens, and over time a mountain forms. Then the cycle repeats.
- Two plates collide and push together and land gets pushed upwards to make a mountain or mountains.
- A mountain is formed by molten rock after a volcano has occurred.
- Over time, rock layers form mountains. They build up and form a mountain.
- Mountains are caused by the land shifting and pushing together, causing the land to go up.
- Mountains form by erosion or volcanoes.
- Mountains are formed when the Earth's plates move and one plate is pushed up to form the mountain.
- During an earthquake sometimes it pushes the ground up with such force that mountains are formed.

3. Give each student a copy of the journal cover sheet (**Blackline Master** *Our Dynamic Planet* P. 2). Direct students to insert the journal cover sheet and their pre-assessment into their journal. Explain that they now have one of the most important tools for this investigation into our dynamic planet: their own scientific journal.

Teaching Tip

What form will journals take? Using loose-leaf notebook paper in a thin three-ring binder enables students to add or remove pages easily. On the downside, loose-leaf pages are more easily lost and students must maintain a regular supply of paper. If you prefer to have students keep journals in composition notebooks or laboratory notebooks, have them trim the journal cover sheet to the appropriate size and paste it onto the first page of their notebooks.

4. Divide students into groups. Instruct the groups to discuss the following:
   • ideas we have about our dynamic planet;
   • questions we have about our dynamic planet.
   One member of the group should record his/her group's ideas and questions on a sheet of poster board, poster paper, or butcher paper.

5. Discuss student responses by having each group, in turn, report on its ideas. As groups are responding, build up two important lists (ideas and questions) for everyone to see (on a chalkboard, flip-chart paper, poster board, or an overhead transparency).

6. Direct students to add these "ideas" and "questions" to their journals.

7. This completes the pre-assessment phase. Distribute copies of *Investigating Our Dynamic Planet*.

Assessment Opportunity

The ideas about dynamic Earth that your students offer here will provide you with pre-assessment data. The experiences they describe, and the way in which they are discussed, will alert you to their general level of understanding about these topics. To encourage this, and to provide a record, it may be useful to quickly summarize the main points about earthquakes and volcanic eruptions that emerge from discussion. You could do this on a chalkboard or flip-chart for all to see. This can be displayed as students work through the module, and added to with each new experience. For your own assessment purposes, it will be useful to keep a record of these early indicators of student understanding.

INVESTIGATING OUR DYNAMIC PLANET

The Earth System

The Earth System is a set of systems that work together in making the world we know. Four of these important systems are:

The Atmosphere

This part of the Earth System is made of the mixture of gases that surround the planet.

The Biosphere

This part of the Earth System is made of all living things, including plants, animals, and other organisms.

The Geosphere

This part of the Earth System is made of the crust, mantle, and inner and outer core.

The Hydrosphere

This part of the Earth System is the planet's water, including oceans, lakes, rivers, ground water, ice, and water vapor.

Introducing the Earth System

Understanding the Earth System is an overall goal of the *Investigating Earth Systems* series. The fact that the Earth functions as a whole, and that all its parts operate together in meaningful ways to make the planet work as a single unit, underlies each module. Each module guides the students in considering this fundamental principle.

At the end of every investigation, students are asked to link what they have discovered with ideas about the Earth System. Questions are provided to guide their thinking, and they are asked to write their responses in their journals. They are also reminded on occasion to record the information on an *Earth System Connection* sheet. This sheet will provide a cumulative record of the connections that the students find as they work through the investigations in the module.

Not all the connections between the things they have been investigating and the Earth System will be immediately apparent to your students. They will probably need your help to understand how some of the things they have been investigating connect to the Earth System. However, by the time they complete the *Investigating Earth Systems* modules to the end of eighth grade, they should have a working knowledge of how they and their environment function as a system within a system, within a system...of the Earth System.

The processes occurring on our dynamic planet, including those discussed in this module, affect the different components of the Earth System in many important ways. The water and gases that form our hydrosphere and atmosphere originated deep within the Earth. Volcanic activity early in Earth's history brought the water and gases that formed the primordial seas and atmosphere to the Earth's surface from within, and continues to do so today. Today, volcanic processes occurring along the Earth's mid-ocean ridges in association with sea-floor spreading play a major role in controlling the chemistry and well-being of the Earth's oceans. Plate tectonic activity creates new lithosphere and destroys old, building continents, mountains, and ocean basins in the process. This obviously affects the geosphere strongly, but it also affects the other Earth systems in profound ways. The positions of the continents and major mountain belts strongly influence climate and the distribution of biomes around the world. The position of the continents also strongly affects whether or not there are large ice sheets on Earth. Earthquakes are another important manifestation of plate tectonics, and the impact of earthquakes on humankind can be seen routinely in today's media. Many cities have been built in seismically active areas. Large earthquakes can be devastating to the people living in those areas. Thus, plate tectonics and the associated volcanic and seismic activity strongly affects the geosphere, hydrosphere, atmosphere, and biosphere.

Distribute copies of the *Earth System Connection* sheet (**Blackline Master** *Our Dynamic Planet* I.1) available at the back of this Teacher's Edition. Have the students place the sheets in their journals. The two sheets may also be copied onto 11 x 17 paper.

Explain to the students that at the end of each investigation they will be asked to reflect on how the questions and outcomes of their investigation relate to the Earth System. Tell them that they should enter any new connections that they discover on the *Earth System Connection* sheet. Encourage them to also include connections that they have made on their own. That is, they should not limit their entries to just those suggested in the **Thinking about the Earth System** questions in **Review and Reflect**. Use the **Review and Reflect** time to direct students' attention to how local issues relate to the questions they have been investigating. By the end of the module, students should have as complete an *Earth System Connection* sheet to *Our Dynamic Planet* as possible.

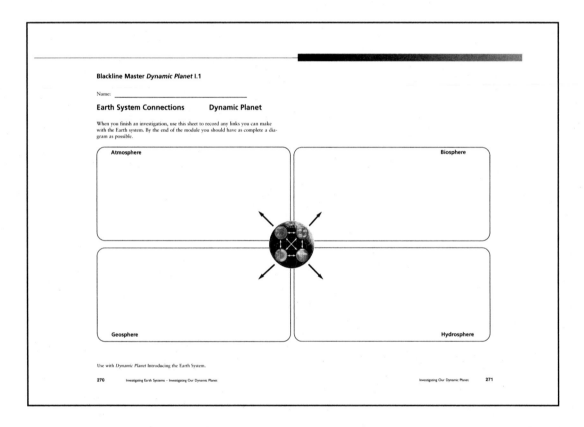

(can be photocopied onto 11x17 paper)

Illustration by Dennis Falcon

INVESTIGATING OUR DYNAMIC PLANET

Introducing Inquiry Processes

When geologists and other scientists investigate the world, they use a set of inquiry processes. Using these processes is very important. They ensure that the research is valid and reliable. In your investigations, you will use these same processes. In this way, you will become a scientist, doing what scientists do. Understanding inquiry processes will help you to investigate questions and solve problems in an orderly way. You will also use inquiry processes in high school, in college, and in your work.

During this module, you will learn when, and how, to use these inquiry processes. Use the chart below as a reference about the inquiry processes.

Inquiry Processes:	How scientists use these processes	
Explore questions to answer by inquiry	Scientists usually form a question to investigate after first looking at what is known about a scientific idea. Sometimes they predict the most likely answer to a question. They base this prediction on what they already know to be true.	**Explore Questions**
Design an investigation	To make sure that the way they test ideas is fair, scientists think very carefully about the design of their investigations. They do this to make sure that the results will be valid and reliable.	**Design Investigations**
Conduct an investigation	After scientists have designed an investigation, they conduct their tests. They observe what happens and record the results. Often, they repeat a test several times to ensure reliable results.	**Conduct Investigations**
Collect and review data using tools	Scientists collect information (data) from their tests. The data may be numerical (numbers), or verbal (words). To collect and manage data, scientists use tools such as computers, calculators, tables, charts, and graphs.	**Collect & Review**
Use evidence to develop ideas	Evidence is very important for scientists. Just as in a court case, it is proven evidence that counts. Scientists look at the evidence other scientists have collected, as well as the evidence they have collected themselves.	**Evidence for Ideas**
Consider evidence for explanations	Finding strong evidence does not always provide the complete answer to a scientific question. Scientists look for likely explanations by studying patterns and relationships within the evidence.	**Consider Evidence**
Seek alternative explanations	Sometimes, the evidence available is not clear or can be interpreted in other ways. If this is so, scientists look for different ways of explaining the evidence. This may lead to a new idea or question to investigate.	**Seek Alternatives**
Show evidence & reasons to others	Scientists communicate their findings to other scientists to see if they agree. Other scientists may then try to repeat the investigation to validate the results.	**Show Evidence**
Use mathematics for science inquiry	Scientists use mathematics in their investigations. Accurate measurement, with suitable units is very important for both collecting and analyzing data. Data often consist of numbers and calculations.	**Use Mathematics**

Introducing Inquiry Processes

Inquiry is at the heart of *Investigating Earth Systems*. That is why each module title begins with the word "Investigating." In the National Science Education Standards, inquiry is the first content standard. (See **Science as Inquiry** on page 8 of this Teacher's Edition.)

Inquiry depends upon active student participation. It is very important to remind students of the steps in the inquiry process as they perform them. Icons that correspond to the nine major components of inquiry appear in the margins of this Teacher's Edition. They point out opportunities to teach and assess inquiry understandings and abilities. Stress the importance of inquiry processes as they occur in your investigations. Provoke students to think about *why* these processes are important. Collecting good data, using evidence, considering alternative explanations, showing evidence to others, and using mathematics are all essential to *IES*. Use examples to demonstrate these processes whenever possible.

At the end of every investigation, students are asked to reflect upon their thinking about scientific inquiry. Refer students to the list of inquiry processes as they answer these questions.

Teaching Tip

If the reading level of the descriptions of inquiry processes is too advanced for some students, you could provide them with illustrations or examples of each of the processes. You may wish to provide students with a copy of the inquiry processes to include in their journals (**Blackline Master** *Our Dynamic Planet* I.2).

Introducing Our Dynamic Planet

Have you ever seen a volcano erupting?

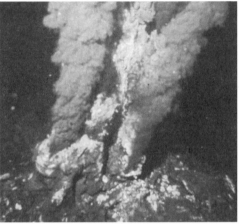

Have you ever heard about hydrothermal vents on the floor of the ocean?

Have you ever wondered how mountains form?

Have you ever seen the effect of an earthquake on a community?

Introducing Our Dynamic Planet

Use this introduction to the module to set your students' investigations into a meaningful context.

This is an opportunity for students to offer some of their own experiences with processes and events that stem from plate tectonics in a general discussion, using the photos and questions as prompts. Some students may be able to cite experiences additional to those asked for here. Encourage a wide-ranging discussion.

Because your students have just spent time in the pre-assessment phase of this module thinking about and discussing what they already know about earthquakes, volcanoes, and mountains, it probably is not necessary to have them complete another journal entry. They will be anxious to get to work on their investigations.

You may want to quickly summarize the main points that emerge from the discussion. You could do this on a chalkboard, flip-chart, or an overhead transparency. For your own assessment purposes, it will be useful to keep a record of these early indicators of student understanding.

About the Photos

The upper left photograph shows a stream of lava flowing on the big island of Hawaii. It may surprise you to know that in the grand scheme of volcanoes, this is a very small eruption! Yet as can be seen, molten rock and fragments of rock are thrown upward in the eruption, and lava flows down the side of the volcano. Each presents a unique kind of hazard to communities and must be investigated and prepared for accordingly.

The upper right photograph shows a hot spring (also known as a hydrothermal vent) on a mid-ocean ridge. Water temperatures greater than 350°C have been recorded at these vents.

The lower left photograph shows a rugged mountain range that has been uplifted by internal forces in the Earth and is now in the process of being worn down, slowly, by erosion.

The lower right photograph is a side view of support-column failure and the collapsed upper deck of the Cypress viaduct and shows just one aspect of the destruction that a major earthquake can cause. The earthquake is known as the October 17, 1989, Loma Prieta California, Earthquake, which occurred during the World Series.

More photos?

INVESTIGATING OUR DYNAMIC PLANET

Why Is Our Dynamic Planet Important?

Dynamic means powerful or active. Our dynamic planet is a powerful, active, ever-changing planet. Powerful events like earthquakes and volcanic eruptions have been happening since the Earth formed, over 4.5 billion years ago. Every feature of our planet changes, on time

scales that range from minutes to millions of years. The deepest oceans, the highest mountain peaks—all represent but a page in the volume of Earth's history. Mountains have been destroyed, recycled, and reborn. Oceans have risen and fallen.

These processes are still at work today. Knowledge about present-day volcanic eruptions and earthquakes give clues about the past. Rocks, landforms, and fossils also provide evidence of a long and varied history of the Earth. Knowing about our dynamic planet will help you to understand the past and prepare for the future.

What Will You Investigate?

You will look for evidence and help solve some of the puzzles surrounding Earth processes. Here are some of the things that you will investigate:

• how scientists make and use models;
• what the inside of the Earth is like;
• how the Earth's surface moves;
• how mountains form;
• what causes earthquakes and volcanic eruptions.

You will need to practice your problem-solving skills. You will also need to be good observers and recorders, as you work together with other members of your class.

In the last investigation you will have a chance to apply all that you have learned about our dynamic planet. You will investigate a natural hazard in depth. Then you will design a brochure to provide information to residents of a community on how to prepare for, and protect against, natural disasters.

Why is Our Dynamic Planet Important?

Read (or have a student read) this section aloud. This introduction gives information from which students can conduct their own investigations throughout the module. You might want to start by having your students read this section carefully, then discuss it in their groups. During their discussions, they can write down a series of questions they may have about dynamic Earth events, or about the investigation itself. You can then have a discussion with your students about their questions.

Students will be familiar with earthquakes and volcanic eruptions. Most will think of them as dramatic, instantaneous, and often unpredictable events. Clearly, students will readily acknowledge that we live on a dynamic planet. This module will *deepen* students' understandings of processes and events on our dynamic planet. For example, students will explore the causes of earthquakes and volcanoes, and they will come to see earthquakes and volcanoes as the provider of "clues" about the interior of the Earth and the history of our planet. Students will also be asked to think about changes on our planet that take place on vast time scales, and they must therefore begin to think of earthquakes and volcanoes as events that occur as part of a process (plate tectonics) that has operated for billions of years. Students will also explore how slow rates of change (e.g., plate motions of 5 cm per year) can produce great changes (e.g., build mountains, separate continents, and form oceans) when they operate over long periods of time (millions of years). These concepts, emphasized in "The Big Picture" that appears on page P72 of the student text, may be abstract and difficult for students to grasp. These ideas are hinted at on page Pxii, and you might consider focusing students' attention to the connections between these ideas and The Big Picture.

About the Photo

The photograph of an eruption of lava illustrates the dynamic nature of our planet. Lava erupts at temperatures greater than 900°C. In the mid-nineteenth century, some geologists inferred from the eruption of lava and the increase in temperature with depth in mines that the entire Earth must be molten below a depth of about 50 km. Scientists used mathematical models of the Earth's rigidity (how the Earth behaves with respect to the gravitational attraction of the Moon) to argue against the "fluid Earth" model and claimed that the Earth behaves like a rigid solid. *Investigating Our Dynamic Planet* will challenge your students to improve their models (their understandings) of our Earth's interior and the forces that make our planet dynamic.

What Will You Investigate?

It is important for students to get a sense of where they are headed in the module. They need your help to connect what seem like unrelated investigations into a cohesive network of ideas. Reviewing this section of the introduction is another step toward constructing a conceptual framework of The Big Picture as it is explored in *Investigating Our Dynamic Planet* (see page P72 of the student text). This framework includes five main concepts:

- how scientists make and use models;
- what the inside of the Earth is like;
- how the Earth's surface moves;
- how mountains form;
- what causes earthquakes and volcanic eruptions.

This would be a good time to review with students the titles of the activities in the Table of Contents. Ask students to explain how the titles of the activities relate to the descriptions in **What Will You Investigate?**

Discussing **Investigation 7** will help students to understand the overall goal of the module. In this final investigation, students use all they have learned to create a brochure that educates the members of a community located near an earthquake or volcano hazard site. Introduce students to the evaluation rubrics so that they can see how their work will be assessed. Sample rubrics are included in the back of this Teacher's Edition.

INVESTIGATION 1: GATHERING EVIDENCE AND MODELING

Background Information

The deep interior of the Earth is inaccessible. Even the deepest boreholes do not penetrate the Earth's mantle below the thin crust. The only direct evidence geoscientists have about the composition of the mantle is from chunks of mantle rock that come to the surface along with basaltic lavas. Basalt is a dark-colored volcanic rock that is produced when the Earth's mantle is partially melted. Much of the volcanic rock that constitutes the ocean crust, ocean islands like the Hawaiian Islands, and parts of the continents is basalt. Sometimes, when the basaltic lava is erupted it contains fragments of the mantle that were entrained with it as the magma (lava that has not yet erupted is called magma) ascended through the mantle into the volcano. These entrained rocks, called xenoliths ("foreign rocks"), are composed mainly of the mineral olivine. This mineral, which you may know as peridot, is thought to be a major constituent of the rock in the upper mantle. The trouble with these xenoliths is that there is no guarantee that their composition is representative of even the upper part of the mantle. Because of this extreme scarcity of direct evidence for the composition of the mantle (to say nothing of the core), geoscientists have had to rely upon various lines of indirect evidence to guide their thinking. The most important guide to the nature of the Earth's interior has come from interpretation of how seismic waves pass through the Earth. That is the subject of **Investigation 2**.

In **Investigation 1** your students will learn much about the nature of models. Geoscientists have been developing models of the Earth's interior for well over 100 years. The "mystery bag" in this activity is a good analogy for the problem of building models of the Earth's interior. Such models could perhaps best be described as conceptual models. These conceptual models of the Earth are guided by various lines of observational evidence, along with certain theoretical considerations derived from physics and chemistry of rocks and minerals.

Before the end of the nineteenth century, geoscientists had only a dim conception of the interior of the Earth. The size and mass of the Earth had been measured quite some time before, so the average density of the Earth was known. That average density is much greater than the density of typical crustal rocks, so it was concluded that the deep interior must be denser than average. The existence of the Earth's magnetic field also gave the clue that the core consists at least partly of iron.

As the world network of seismic (earthquake) observatories was developed around the beginning of the twentieth century, the way was opened to construction of much more detailed and realistic models of the Earth's structure (see the diagram on page P20 of the students' text). The existence of a sharp boundary between the crust and the mantle, and between the mantle and the core, was recognized early on. In the early decades of the twentieth century, the existence of a liquid outer core and a solid inner core was discovered by means of some fancy detective work on the basis of seismic (earthquake) waves. Much more recently, as the number of seismic stations covering the Earth's surface has increased, a technique called seismic tomography has allowed geoscientists to image the geometry and position lithospheric plates as they are subducted into the mantle. Tomography uses many different sources of seismic energy to image structures in three dimensions, and it is analogous to remote sounding techniques like the CAT scans (among others) used in medicine. Nowadays, understanding of the deep structure of the Earth is increasing rapidly, thanks to such techniques.

More Information…on the Web
Go to the *Investigating Earth Systems* web site www.agiweb.org/ies for links to a variety of other web sites that will help you deepen your understanding of content and prepare you to teach this module

Investigation Overview

Investigation 1 presents students with a "mystery bag," and challenges them to use their senses to construct a model of its contents. Students begin with the sense of smell, then use their sense of hearing, and finally use their sense of touch. Students discuss their models and revise their models as they gather new evidence. Finally students are asked to speculate about other kinds of "tests" or methods of gathering information about the contents. They read in the **Digging Deeper** section about the different kinds of models that scientists use, and the different ways that scientists use the terms "model" and "modeling." This investigation sets the stage for **Investigation 2**, in which students explore the evidence that scientists have used to construct a model of something that they cannot "see," the Earth's interior.

Goals and Objectives

As a result of this investigation, students will develop a better understanding of models and will improve their ability to make and record observations.

Science Content Objectives

Students will collect evidence that:
1. Something that cannot be seen can be modeled.
2. Scientists use the words "model" and "modeling" in different ways.
3. Models can be physical, conceptual, mathematical, or numerical.
4. Models are often revised as new evidence is gathered.

Inquiry Process Skills

Students will:
1. Make observations using the senses.
2. Record observations in a systematic way.
3. Make a model of what they cannot see.
4. Revise models based on additional observations.
5. Communicate observations and findings to others.

Connections to Standards and Benchmarks

In **Investigation 1**, students will make inferences about the contents of a mystery bag, and produce a model of its contents. Students will revise these models as they gather new evidence using different senses. These observations will start them on the road to understanding the National Science Education Standards and AAAS Benchmarks shown below.

NSES Links

- Evidence consists of observations and data on which to base scientific explanations.

- Models are tentative schemes or structures that correspond to real objects, events, or classes of events, and that have explanatory power. Models help scientists and engineers understand how things work. Models take many forms, including physical objects, plans, mental constructs, mathematical equations, and computer simulations.

AAAS Link

- Models are often used to think about processes that happen too slowly, too quickly, or on too small a scale to be observed directly, or that are too vast to be changed deliberately, or that are potentially dangerous.

Preparation and Materials Needed

Preparation

Prior to this investigation, you will need to prepare a mystery bag for each group. Here is how to do this:

Bag contents: *lemon peel*
- one apple
- one small potato
- several candies (hard and soft)
- several coins

Put the candies and coins in a thin plastic bag; fill the plastic bag halfway with air, and seal it with a twist-tie (alternatively, use a ziplock-style bag). Place the plastic bag and other items inside each brown paper grocery bag, roll the top of the bag, and staple or tape it shut. Just before the investigation starts (see **Step 3**), and out of the sight of students, rub lemon peels or put pure orange extract on each paper bag to give it the scent of citrus. NOTE: If you are intending to use this with more than one class, you will need to refresh the bags each time, or use lemon essential oil as an alternative. If you are unable to scent the bag out of sight of the students (self-contained classroom, consecutive class periods, etc.), that's okay. Students can still use their sense of smell to try to detect what is in the bag. You may also need to "double bag" the items to extend their life.

If your students are not used to working in small collaborative groups, spend some time helping them understand how to work together. Keep in mind that some students may find it difficult to work in a group (some prefer to work alone). Help students understand that collaborating means working together and that this is an important part of scientific inquiry. Sample rubrics for evaluating group participation are provided in the appendix to this Teacher's Edition (see **Group Participation Evaluation Sheets I and II**). Discussing the criteria will help to reinforce the importance of individual accountability and cooperation.

Materials

For each student:
- white paper
- crayons, markers or colored pencils

For each group:
Mystery bag (brown paper grocery sack) with objects sealed inside:
- one apple
- one small potato
- several candies (hard and soft)
- several coins
- a thin plastic bag with twist-tie

The teacher will need:
- orange peels, orange extract, lemon extract, or some other fragrance
  to put a scent on the paper bag

Investigation 1:

Gathering Evidence and Modeling

Key Question

Before you begin, first think about this key question.

How do you make a model of something that you cannot see?

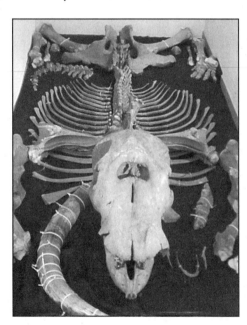

Think about what you know about models. What are some models of things or processes that cannot be seen with the naked eye? How do you think these models were constructed?

Discuss the key question with your group and your class. Record your thoughts in your journal. Be prepared to share your thinking with the rest of the class.

Materials Needed

For this investigation your group will need:

- double-bagged, brown paper bag with "mystery objects" sealed inside

- colored pencils

Investigate

1. Several kinds of objects have been placed into a brown paper "mystery bag," and it has been sealed shut. You will use three senses (smell, hearing, and touch) to gather data about the bag's contents. Then you will design new tests to get more information.

Key Question

The **Key Question** is a brief warm-up activity to elicit students' ideas about models. The responses to the question will alert you to the level of understanding of modeling these students bring to this investigation. When conducting this warm-up, promote thinking and sharing of ideas. Avoid seeking closure (i.e., the "right answer"). Closure will come through inquiry, reading the text (**Digging Deeper**), discussing the ideas (lecture), and reflecting on what was learned at the end of the investigation. Make students feel comfortable sharing their ideas by avoiding commentary on the correctness of responses.

Write the **Key Question** on the chalkboard or on an overhead transparency. Have the students record their answers in their journals. Tell them to think about and answer the question individually. Tell them to write as much as they know and to provide as much detail as possible in their responses. Emphasize that the current date and the prompt (the question itself, a meaningful heading, etc.) should be included in journal entries.

[handwritten: QD pair/share]

[handwritten: Alt: put out series of models, ask them to ID what they have in common.]

Student Conceptions about Models

When asked to think about a model of something that cannot be seen, students suggest physical models of the interior of something that cannot be seen directly, or which no longer exists. For example, students commonly suggest a plastic model of the inside of a person (organs, bones, muscles) because we can't see through skin. They will also suggest physical models of dinosaurs, which are made on the basis of fossils, even though no one has ever seen a dinosaur. Some students will suggest how the use of x-rays and sonograms/ultrasounds can be used to produce images of the inside of the body. Whether or not these are models in the true sense of the word can be debated, but they help to illustrate how to render image of something that cannot be seen which is the point of the exercise.

Answer for the Teacher Only

Models come in many different forms, including mathematical models (computer-based models of global climate change), conceptual models (models of the behavior of the structure and behavior of subatomic particles, which are too small to be imaged, or of the formation of the solar system, which occurred too long ago to have been witnessed). The inside of the Earth cannot be seen. Scientists have to gather evidence about the Earth to find out what materials make up the interior. A first step is to make a model. Over many years, scientists have pieced together enough evidence and have revised their models of the interior of the Earth many times. As you work through this module you will learn more about this evidence.

Assessment Tool

Key Question Evaluation Sheet

Use this evaluation sheet to help students understand and internalize basic expectations for the warm-up activity. The **Key Question Evaluation Sheet** emphasizes that you want to see evidence of prior knowledge and that students should communicate their thinking clearly. You will not likely have time to apply this assessment every time students complete a warm-up activity; yet, in order to ensure that students value committing their initial conceptions to paper and taking the warm up seriously, you should always remind them of the criteria. When time permits, use this evaluation sheet as a spot check on the quality of their work. As with any assessment tool used in *IES*, the assessment instrument should be provided to students and discussed *before* they complete a task. This ensures that they have a clear understanding of your expectations for their work.

Investigate

Teaching Suggestions and Sample Answers

It is important that your students understand that what they are about to do represents a scientific approach to gathering evidence. Help them to review the ideas given here. You might also point out to students that the data they will collect is the evidence

Assessment Tools

Journal Entry-Evaluation Sheet   pg 301

Use this sheet as a general guideline for assessing student journals, adapting it to your classroom if desired. You should give the **Journal Entry-Evaluation Sheet** to students early in the module, discuss it with them, and use it to provide clear and prompt feedback.

Journal Entry-Checklist   pg 302

Use this checklist for quickly checking the quality and completeness of journal entries. You can assign a value to each criterion, or assign a "+" or "-" for each category, which you can translate into points later.

INVESTIGATING OUR DYNAMIC PLANET

In your journal, write your research question: "What are the contents of the Mystery Bag?" Underneath your research question, write your prediction.

Conduct Investigations

WHAT ARE THE CONTENTS OF THE MYSTERY BAG?

SMELL	HEARING
Model	**Model**
Evidence	**Evidence**

TOUCH	FURTHER TESTS
Model	
Evidence	

⚠ Do not use taste. The items in the bag are safe to smell. Follow safe procedures when smelling unfamiliar materials.

2. Divide a sheet of paper into four equal sections. Label the four sections as follows:

• Smell

• Hearing

• Touch

• Further Tests

Divide the sections for smell, hearing, and touch, into two sections: model and evidence. In the evidence section you will write observations that support your ideas about what you think is in the bag. In the model section you will draw a model of what you think is in the bag.

1. Students should record their ideas about what they think might be in the mystery bag. The first hypothesis will be based upon the sense of sight, because the size of the bag will constrain the potential contents to some extent. Students who get near the bag often notice a particular odor and use their sense of smell to inform their thinking.

2. Provide students with a copy of **Blackline Master** *Our Dynamic Planet* 1.1, What are the Contents of the Mystery Bag? Be sure to familiarize yourself with these ahead of time. Remind students how the senses (sight, touch, smell, sound, and in some safe instances, taste) can all be used to make observations. Stress also that observations can provide evidence that can then be evaluated and analyzed to provide explanations for scientific inquiry questions. Students need to understand that accurate observations are crucial in this respect and that this often means taking a systematic approach. Without this, it is easy to overlook something that might be important to accuracy, and therefore the validity of the observation.

Investigation 1: Gathering Evidence and Modeling

Collect & Review

3. The first sense you will use is smell. Your teacher will place the bag on the center of your table. Smell the bag without touching it.

   a) Record your ideas about what is in the bag in the "model" section of the square, using pictures.

Evidence for Ideas

   b) In the other section of the box, record your evidence. This is like a justification, or explanation, of why you drew the picture the way you did.

Collect & Review

4. Next, you will use your sense of hearing. One member of your group should pick up the bag and shake it, while walking around to each group member.

   a) Record your model(s) in the correct square, using pictures.

Evidence for Ideas

   b) Record your evidence.

Collect & Review

5. Next, you will use your sense of touch, using care not to open or damage the bag, or the bag's contents. Allow each group member to touch the bag.

   a) Record your model(s) in the correct square, using pictures.

Evidence for Ideas

   b) Record your evidence.

Show Evidence

6. Discuss your models and evidence in your group.

   When you have come to an agreement, select a group member to share one of your group's models and evidence with the rest of the class.

7. After discussing your class's findings, compose a list of further tests you could perform to gather more evidence. These do not have to be tests that you will actually try in the classroom, although they could be. The only rule is that you may not look inside the bag.

   a) Write your group's ideas under "Further Tests" in your journal.

   b) Share your ideas in a group discussion.

Inquiry

Using Evidence

Evidence is very important for scientists. They can use evidence to develop conceptual models (what they think something they cannot see might look like). Evidence comes from observation and data. In this investigation you can use the data you collect from your observations as evidence to develop a conceptual model of what is in the brown paper bag.

Have your teacher check your tests for safety if you plan on trying any of them. Wash your hands after the activity.

P
3

3. As you place the bag on students' tables, remind students not to touch the bag. You can cite this as an example of making quality observations. Explain that by touching the bag they may interfere with the validity of the test; for example, odors may be transferred from their hands onto the bag.

   a) You may wish to have students sketch what they think is in the bag. The first models will depend upon how you have scented the bag. If you've rubbed an orange onto the bag, oranges will appear in students' models.

   b) Remind them to also note the evidence that supports their inference about the contents of the bag. Following the example described above, a sentence like "The bag smelled like oranges" would be an appropriate citation of evidence.

4. Students may not immediately realize that observations of sound can be important scientific data. Provide an example, like a doctor listening to a patient's heartbeat with a stethoscope. You will need to ensure that the classroom is relatively quiet for students to make good sound observations. Be sensitive to students with impaired hearing. They will need to team with students who do not have this disadvantage.

   a) – b) Encourage students to keep accurate records during the investigation and to always record their evidence.

5. Remind students that we often use our sense of touch to identify objects, especially when they are hidden from view, like items in coat pockets or small things in a school backpack. Ask them to think about how they would find their way around a room if it were suddenly plunged into darkness because of a power outage at night. Help students to be systematic about using touch with the bag.

   a) – b) Remind students about the importance of recording their model as well as the new evidence that supports their model.

6. Circulate among student groups. Check to see that students have recorded their models, as well as their evidence. Glance at student work as you go, and identify several students whose models differ widely. Note these so that you can bring these differences out in the class discussion that follows.

   In a class discussion, have students share their models. Select at least one volunteer from each table. Ask students what they think is in the bag, and reinforce the important role of evidence by asking them to describe the evidence for their model.

7. Your students may come up with some creative ideas here. Encourage creative thinking on what other tests might be used to help clarify the contents of the bag.

   a) – b) Students often suggest drilling a core sample, taking an x-ray, or measuring the mass of the bag.

Teaching Tip

The most often asked question at the completion of this activity is "Can we open the bag?" You might reply to students "Well, if we want to know if our model of the inside of the Earth is right, can we open it up?" Allowing students to open the bag creates two problems. First, the mystery and the inquiry end, which is the last thing that you want to promote in a science classroom (better to leave students wondering, as scientists are left to wonder about their models). You can also expect that some students will have difficulty keeping this mystery a secret with following classes.

Assessment Tool

Investigation Journal Entry-Evaluation Sheet

Use this sheet to help students learn the basic expectations for journal entries that feature the write-up of investigations. It provides a variety of criteria that both you and your students can use to ensure that their work meets the highest possible standards and expectations. Adapt this sheet so that it is appropriate for your classroom or tailor the sheet to suit a particular investigation.

NOTES

INVESTIGATING OUR D

Digging Deeper

Evidence
for Ideas

As You Read...
Think about:
1. *What is the difference between a physical model and a conceptual model?*
2. *What is the difference between a hypothesis and a model?*
3. *How does mathematics help in developing models?*
4. *How do computers help in developing models?*

MODELS

In science there are many kinds of models. Scientists use the words "model" and "modeling" in many different ways.

Physical Models

Physical models are structures that scientists build to represent something else. This kind of model is probably what would pop into your mind first if somebody asked you what a model is. The simplest kind of physical model is just a small-scale structure of what a much larger item looks like, or once looked like but no longer exists. The dinosaurs you might see in a museum are models built by paleontologists (scientists who study fossils and ancient life). They collect the fossilized bones and then make plaster casts of them. They try to fit the bones together in the most realistic way. Then they try to imagine what the flesh and skin might have looked like.

Digging Deeper

This section provides text, illustrations, and photographs that give students greater insight into four different kinds of models. You may wish to assign the **As You Read** questions as homework to help students focus on the major ideas in the text.

About the Photo

This is an example of a physical model of the solar system. As simple and conventional as this model appears, you can point out to students that it has changed significantly over the last four hundred years (before Copernicus and Galileo, most people accepted that the Sun revolved around the Earth!). This photograph is also an opportunity to point out that all models have limitations. For example, in this physical model, Pluto is on the same orbital plane as the rest of the planets in the solar system. In reality, Pluto's orbital plane is inclined relative to those of the other planets. This would be difficult to accomplish in a physical model because the steel bar that attaches Pluto to the model would have to cross those of the other planets. Another limitation to this physical model is that the relative sizes of planets and distances between planets are not shown to scale. For example, the diameter of the Sun is actually 110 times that of Earth, and Jupiter is more than five times as far away from the Sun than is the Earth.

As You Read...

Think about:

1. Physical models are real structures that scientists build to represent something else. An example is a dinosaur skeleton. Conceptual models are models that scientists develop in their minds, using what has been observed and knowledge about physical and chemical laws of nature.

2. A hypothesis and a model have common elements, such as being based upon what one knows about a system or process. A hypothesis is narrower in scope; it deals with fewer aspects of simpler natural systems.

3. If a scientist can represent a natural process with a mathematical equation, then he or she can use the equation to predict how the process will change by changing a value in the equation and doing the calculation. For example, if scientists represent force as the product of mass and acceleration, then by substituting different values for the mass and making calculations the scientist can predict how the force will vary with changing mass.

4. Computers enable scientists to handle very complex mathematical models, models that would take far too long to do the calculations by hand.

Assessment Opportunity

You may wish to rephrase selected questions from the **As You Read** section into multiple choice or "true/false" format to use as a quiz. Use this quiz to assess student understanding and as a motivational tool to ensure that students complete the reading assignment and comprehend the main ideas.

NOTES

Other physical models simulate some event or process in nature. Very large-scale natural processes like river-flow or ocean waves are difficult to study in the outdoors. Scientists build tanks, channels, or basins to reproduce the processes in a laboratory. Sometimes they are able to adjust the conditions (things like speed of water-flow, or the behavior of water waves). Then what they observe in the model represents what happens in the outdoors. Even if they are not able to do that, then at least they are able to get some valuable qualitative data just by watching what happens in the model.

Conceptual Models

Conceptual models are models that scientists develop in their minds. Scientists often try to develop a concept about how some process works in nature. The basis for the concept is what has already been observed about the process, together with what the scientist knows about basic physical or chemical laws. A conceptual model is a bit like a hypothesis. It is usually broader than a hypothesis, however, because it deals with many things about a complicated natural system. A good example of a conceptual model is the picture scientists have about the nature of atoms. You probably know that atoms consist of a nucleus, and electrons that orbit around the nucleus. The nucleus consists of protons and neutrons. That's the simplest conceptual model of an atom. With the discovery of even more elementary particles, that original conceptual model of an atom has been enlarged and extended.

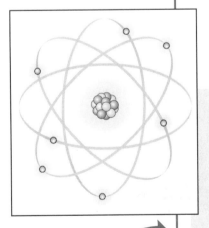

Inquiry

Hypotheses

A hypothesis is a testable statement or idea about how something works. It is based on what you think that you know or understand already. A hypothesis is never a guess. You test a hypothesis by comparing it to observations or data that already exist or that can be gathered in the future. A hypothesis forms the basis for making a prediction, and is used to design an experiment or observation to find out more about a scientific idea or question. Guesses can be useful in science, but they are not h

About the Illustration

The diagram shows a model of the atoms with three basic components: neutrons and protons in the nucleus, and electrons orbiting the nucleus. For centuries, scientists have produced and revised conceptual models of the nature of matter. A simpler and earlier conceptual model than the one shown was one without a nucleus. A modern, more complex model details more than a hundred subatomic particles!

Mathematical Models

Sometimes, scientists are able to write mathematical equations that describe how some process works. The equations express basic physical and chemical laws. Then they solve the equations, in the same way that students in a math class solve equations. The solutions to the equations tell how the process will work, under a variety of conditions. This allows scientists to predict what will happen—which is one of the important things scientists try to do.

Numerical Models

Sometimes, scientists are able to develop a mathematical model, but the model would take too long to fully test by hand. That's where high-speed computers come in. The equations are programmed into the computer. The computer can then compute the model thousands of times. In this way it can quickly simulate what happens either over time, or with changing conditions. As computing power has grown in recent years, the ability of computers to handle numerical models has gotten much greater. Several groups of climatologists (scientists who study global climate) have developed numerical models to study how the Earth's climate might change in the coming decades. As more and more is known about processes of climate, the models are continually being refined.

About the Illustration

Meteorologists use complex mathematical models to forecast weather.

Making Connections...*with Mathematics*

The **Digging Deeper** reading section describes numerical models. You might consider discussing this description with your students' math teachers to identify connections between the math and science class.

Review and Reflect

Review

1. Look again at the **Key Question** for this investigation. In your journal, write down what you have learned from your investigation that provides answers.

2. Describe how and why your third model (using touch) is a better model than your first model (using smell).

3. Of the four kinds of models described in the **Digging Deeper** reading section, how would you classify your model of the mystery bag? Explain your answer.

Reflect

4. Explain how time and cost considerations might affect the process of making a model.

5. How does technology affect the ability of scientists to develop models?

6. Describe a situation in which a model is made of something that cannot be observed directly. Use an example that has not been given in this investigation.

Thinking about the Earth System

7. The modeling that you did with the mystery bag can be connected to the Earth's major systems. Provide a connection to each of the Earth systems in this investigation. Remember to write connections, as you find them, on the *Earth System Connection* sheet.

Thinking about Scientific Inquiry

8. What role did evidence play in your group's models?

9. How was communicating findings to others an important scientific process in this activity?

10. In what ways would the process of making a model of the Earth's interior be like the mystery bag activity?

Review and Reflect

Review

When your students began this investigation, they may not have thought about how useful models can be in showing how we think something "works," how scientists have developed different kinds of models for different purposes, and that models can be revised and improved as we obtain new evidence. These ideas are central to the nature of science. Use the questions below to help students to think more deeply about these core ideas explored in the activity and reading in this investigation. A solid understanding of models will make it easier for students to understand how models of the Earth's interior have changed over time. This will be explored in **Investigation 2**.

The answers provided below are for you, the teacher. It is not expected that your students will answer with the same level of sophistication. Use your knowledge of the students as well as the standards set by your school district to decide what answers you will accept. In student answers, look for evidence of an understanding of the processes involved, as well as for any misconceptions that still remain. Encourage students to express their ideas clearly, and to use correct science terminology where appropriate.

1. Student answers will vary but should include some reference to the fact that we can use different ways of gathering evidence (different senses were used in this investigation), think about the evidence we gather, and use this evidence to hypothesize about what we cannot see.

2. The smell of the bag suggested that the bag contained a particular object (an orange or lemon, depending upon the kind of scent was put on the bag). The opportunity to touch the bag allowed students to reject that hypothesis. It also allowed students to gather evidence about objects that they could not smell, like the candies and coins in the plastic bag, the plastic bag itself, the potato, and the apple.

3. Students should note that their models are most similar to conceptual models because conceptual models are models that they developed in their minds. The model of the mystery bag was not a physical model (although it was written down on paper), and students did not represent the mystery bag through a mathematical equation or use computations to develop the model.

Reflect

It is very important that your students be given adequate time to review and reflect on what they have done and understood in this investigation. Ensure that all students think about and discuss the questions listed here. Be on the lookout for any misunderstandings and, where necessary, help students to clarify their ideas.

4. A physical model, especially one that simulates events or processes in nature, is designed to be small enough to model the process at an acceptable cost. Numerical models save money because computers can process complex mathematical equations rapidly; it would cost far more to pay people to do the calculations by hand.

5. Technology allows us to gather new evidence that we can use to refine or improve our models. Technology, particularly high-speed computers, allows us to quickly process large amounts of evidence (data) in numerical models.

6. Answers will vary. Examples include a model of how life evolved on Earth, a model of a chemical reaction, and a model of deep ocean currents.

Thinking about the Earth System

It is very important that students begin to relate what they are studying to the wider idea of the Earth System. This is a complex and largely inferred set of concepts that students cannot easily understand from direct observation. Remember, the goal is that students will have a working understanding of the Earth system by the time they complete eighth grade. Although it can be taught as a piece of information, true understanding is largely dependent upon comprehending how numerous specific Earth science concepts connect with the idea of the Earth as a system. Be sure to spend some time helping students make what connections they can between the focus of their investigations and this wider aspect.

7. Answers will vary. Because this investigation focused on the nature of science and how models are created and revised, you should not expect significant detail in student responses at this point in the module. Students are most likely to focus on the geosphere and how we have developed models of the interior of the Earth.

Thinking about Scientific Inquiry

Science as inquiry is a theme that runs through all investigations. Students will need many investigative experiences to grasp the many processes and skills involved with scientific inquiry. This can be taught as a piece of information, but for a solid understanding, students need considerable firsthand experience. They are given many opportunities to think about the connections between their investigations and inquiry processes.

8. Evidence is essential to the formation and revision of models. Different senses produced different kinds of evidence. Each time new evidence was gathered, students had to reevaluate their models in light of the new information.

9. Students should note that discussing their models in their groups, and as a whole class, led them to think about their models (because they had to describe their

thinking). Feedback to their models, or listening to other groups present models and evidence, might have led them to collect more evidence (recheck the bag) and revise their model.

10. Students were not allowed to open the mystery bag, and we are not able to "open the Earth" to see if our models are correct. Making models of the Earth's interior requires evidence, as did the making of models of the mystery bag. One would also expect models of the Earth's interior to be revised as new evidence is gathered, as with the mystery bag activity.

Assessment Tool
Review and Reflect Journal Entry-Evaluation Sheet
Depending upon whether you have students complete the work individually or within a group, use the **Review and Reflect** part of the investigation to assess individual or collective understandings about the concepts and inquiry processes explored. Whatever choice you make, this evaluation sheet provides you with a few general criteria for assessing content and thoroughness of student work. Adapt and modify the sheet to meet your needs. Consider involving students in selecting and modifying the assessment criteria.

Teacher Review

Use this section to reflect on and review the investigation. Keep in mind that your notes here are likely to be especially helpful when you teach this investigation again. Questions listed here are examples only.

Student Achievement

What evidence do you have that all students have met the science content objectives?

_____

Are there any students who need more help in reaching these objectives? If so, how can you provide this? _____

_____

What evidence do you have that all students have demonstrated their understanding of the inquiry processes? _____

_____

Which of these inquiry objectives do your students need to improve upon in future investigations? _____

_____

What evidence do the journal entries contain about what your students learned from this investigation? _____

_____

Planning

How well did this investigation fit into your class time? _____

_____

What changes can you make to improve your planning next time? _____

_____

Guiding and Facilitating Learning

How well did you focus and support inquiry while interacting with students?

_____

What changes can you make to improve classroom management for the next investigation or the next time you teach this investigation? _____

_____

How successful were you in encouraging all students to participate fully in science learning? _____

_____

How did you encourage and model the skills values, and attitudes of scientific inquiry? _____

_____

How did you nurture collaboration among students? _____

_____

Materials and Resources

What challenges did you encounter obtaining or using materials and/or resources needed for the activity? _____

_____

What changes can you make to better obtain and better manage materials and resources next time? _____

_____

Student Evaluation

Describe how you evaluated student progress. What worked well? What needs to be improved? _____

_____

How will you adapt your evaluation methods for next time? _____

_____

Describe how you guided students in self-assessment. _____

_____

Self Evaluation

How would you rate your teaching of this investigation? _____

_____

What advice would you give to a colleague who is planning to teach this investigation? _____

_____

NOTES

Investigating Earth Systems – Investigating Our Dynamic Planet

INVESTIGATION 2: THE INTERIOR OF THE EARTH

Background Information

Stresses in the Earth's Outer Layer

The forces that act in the Earth's rigid outer layer are transmitted over long distances through the rocks. The principle is the same as when you hold the ends of a long rod of wood or metal and try to pull the rod apart, or squeeze it together, or bend it. This results in local internal forces everywhere in the rod, called stresses. The stresses in rocks in the Earth can be compressional (tending to squeeze the material together), extensional (tending to pull the material apart), or shear (tending to cause sideways sliding in the material). The stresses cause the rock to change its shape (expanding, contracting, or bending). This change in shape is called deformation, or strain. As rock is strained, energy is stored in the rock. It's the same as when you compress or stretch a metal spring. You are storing energy in the spring, and that energy shows up as the motion of the spring when you relax your force on it.

Over time, stresses can build up to the point where they exceed the strength of the rock, causing it to break and releasing the stored energy in the form of an earthquake. The energy of strain can accumulate in rocks over decades, centuries, or millennia and be released in seconds or minutes. In this model of how rocks respond to stresses, the rocks deform elastically, in the sense that when the forces are relaxed by the breaking of the rock, the rock tends to spring back to its original shape. This phenomenon is called elastic rebound. The model of elastic rebound is widely accepted by geoscientists as accounting for the behavior of rocks during an earthquake.

In the model of elastic rebound, when the force exceeds the strength of the rock and the rock breaks, the rock masses on either side of the fracture surface slide past each other as they return to their original undeformed shape. The surface along which the sliding takes place is called a fault. As the rock masses slide past each other, the elastic energy that was stored within them is released in the form of waves. The motions of materials as these waves pass through them are what is felt as an earthquake.

The waves generated by an earthquake move outward in all directions away from the place, called the focus, where the earthquake happened. These waves are like water waves, which your students experiment with in **Part A** of this investigation, in the sense that they cause an oscillating motion of the material, and this oscillation is propagated from place to place without the material itself undergoing a permanent change in position. Just as with the water, the rock through which the seismic waves move comes to rest again, in its original position, once the waves have passed by.

In contrast to water waves, seismic waves travel at very high speeds of several kilometers per second through rock. If you want to give your students some idea of how fast such waves move, you might take them outside the school building, place them in a group, walk fifty meters away from them, and make some loud sound, as with a basketball hitting pavement. They will hear the sound slightly later than they see you make the impact. Sound waves are exactly like seismic waves (the compressional kind, not the shear kind), but the waves in the rock move even faster than in the air.

Demo: Silly Puddy

Investigation 2

demo

To a first approximation, the wave fronts of seismic waves have a spherical shape in three dimensions, just as the ripples that are formed on a still water surface when a pebble is dropped into the water have a circular shape in two dimensions. For convenience in visualizing the travels of the waves, however, imaginary curves called rays are constructed to be everywhere perpendicular to the actual wave fronts. When geoscientists show diagrams of the passage of body waves through the Earth, they usually show the waves in the form of the rays rather than the wave fronts.

When a seismic wave reaches a surface of abrupt change in the speed of propagation of the seismic waves, caused by some change in the composition or density of the rock, the rays are bent or redirected at the surface of discontinuity. This is what the students model in **Part C** of this investigation. Another way of visualizing abrupt refraction of waves is simply to stick a pencil into a glass of water. The pencil looks as though it is bent where it enters the water, because the speed of light waves is different in the two media (air and water).

If the speed of movement of the wave fronts were the same everywhere in the Earth, the rays would simply be straight lines extending in all directions from the earthquake focus. However, the speed of movement of the waves varies with depth in the Earth, generally increasing with depth. That causes the rays to be curved in such a way as to be generally concave upward. This phenomenon is another manifestation of refraction. It involves continuous refraction through a slowly varying medium rather than abrupt refraction at a surface of discontinuity.

More Information...on the Web
Go to the *Investigating Earth Systems* web site www.agiweb.org/ies for links to a variety of other web sites that will help you deepen your understanding of content and prepare you to teach this module.

Investigation Overview

Students conduct four interrelated investigations into the evidence that scientists have used to develop a model of the Earth's interior structure and composition. Students begin by exploring the movement of water waves to gain a hands-on perspective on the velocity of waves in a uniform material. Students then use Slinkys® to explore differences in wave velocity and motion between two different kinds of seismic waves. In the third part of the investigation, students explore wave refraction, which happens when a wave crosses a boundary between materials with different densities. Finally, students apply what they have learned about wave motion and refraction to predict what happens when seismic waves reach the Earth's core. These four parts of the investigation provide important experiences needed to help students to understand the text presented in the **Digging Deeper** reading section, which formally presents the evidence that scientists have used to construct and revise models of the Earth's interior.

Goals and Objectives

As a result of this investigation, students will understand some of the evidence used to construct and revise models of the Earth's interior structure, and will understand why we believe that the Earth has a layered structure.

Science Content Objectives

Students will collect evidence that:
1. Waves travel at a certain rate in a uniform material.
2. Waves carry energy with them as they move through a material.
3. Waves are refracted when they enter material of a different density.
4. Compressional waves can travel through solids, liquids, and gases, but shear waves can travel only through solids.
5. The Earth's interior has a layered structure including crust, mantle, and core, which we have inferred from the study of seismic waves.
6. Models can be revised and improved with additional data.

Inquiry Process Skills

Students will:
1. Predict the behavior of waves in water.
2. Conduct fair and objective tests.
3. Measure distance and time.
4. Calculate rate.
5. Make observations using the senses.
6. Record observations.
7. Devise a plan to investigate a question.
8. Collect observational data.
9. Compare data sets.
10. Use test data to interpret a model.
11. Communicate observations and findings to others.

Connections to Standards and Benchmarks

In **Investigation 2**, students investigate key evidence used to develop a model of the Earth's interior—the study of seismic waves. Understanding this evidence and the general structure and properties of the Earth's interior will set the stage for understanding what makes the Earth's interior "dynamic", which will be explored in depth in **Investigation 3**. These observations contribute to developing the National Science Education Standards and AAAS Benchmarks shown below.

NSES Links

The solid Earth is layered with the lithosphere; hot, convecting mantle; and dense, metallic core.

Lithospheric plates on the scales of continents and oceans consistently move at the rate of centimeters per year in response to movements in the mantle. Major geological events, such as earthquakes, volcanic eruptions, and mountain building, result from these plate motions.

AAAS Links

Vibrations in materials set up wavelike disturbances that spread away from the source. Sound and earthquake waves are examples. These and other waves move at different speeds in different materials.

The Earth is mostly rock. Three-fourths of its surface is covered by a relatively thin layer of water (some of it frozen), and the entire planet is surrounded by a relatively thin blanket of air. It is the only body in the solar system that appears able to support life. The other planets have compositions and conditions very different from the Earth's.

The interior of the Earth is hot. Heat flow and movement of material within the Earth cause earthquakes and volcanic eruptions and create mountains and ocean basins. Gas and dust from large volcanoes can change the atmosphere.

Preparation and Materials Needed

Preparation

An important part of preparing for **Part A** of the **Investigation** is to try it ahead of
time to get a handle on the subtleties behind timing the speed of the wave. This will
help you to provide advice to your students. When you look down the flashlight into
the water, you will see a reflection of the flashlight bulb. It will be very easy to see
that reflection move (wiggle) when the wave arrives. It takes a little practice, but
works quite well. If you do not have enough flashlights (one needed for each group),
you might ask for a volunteer from each group to bring one from home. Stopwatches
are needed for students to make accurate measurements. To save time in the activity,
you can mark the dots on the bottom of each pan and have a container of water at
each group's station before class (the first part of **Step 1**).

Part B requires two extra-long Slinkys® per student group. If you do not have a
large collection of these, have groups rotate through this part of the investigation
rather than substitute the regular-length Slinky. This activity does not work well
on a carpeted floor, so find a smooth surface. The activity will take only about five
minutes to complete.

Part C works best when done on a school playground or large flat surface
(gymnasium floor), but you may be able to pull this off in your classroom if you can
move desks and chairs to the side of the room and create enough open space. If you
have a carpeted room or do the activity on a wooden floor, use masking tape rather
than chalk to mark the lines. Photocopy the **Blackline Master** *Our Dynamic Planet*
2.2, Modeling Wave Refraction (one per student) ahead of time. To save time, prepare
the white line and red line ahead of time (**Step 1** on pages P12 and 13). The activity
works very well in demonstrating refraction, but you will need to advise students to
walk in regular steps (if they race to catch up with one another, the refraction effect is
not easy to see), to continue to walk straight ahead after crossing the white line (they
will cross the line at an angle, but should move straight ahead), and to stop exactly
when asked to do so. Use a large protractor (the kind used for chalkboard angles in
math classes) to measure angles formed by the lines of students. **Step 11** of this part
of the investigation requires a clear container filled with water and a long thin stick-
like object (pencil, chopstick, wooden dowel).

For **Part D**, photocopy the **Blackline Master** *Our Dynamic Planet* 2.3, Refraction of
Earthquake Waves for each student. Photocopying this on two sides of the paper will
give students an opportunity to practice on one side (almost always bound to have
mistakes and erasures) and complete their final diagram on the other. Students will
need a transparent straightedge to draw the lines of refraction.

Suggested Materials

Part A
Materials Needed (per student group):
For this part of the investigation your group will need:

handwritten: – sand table container ?

- large, flat-bottomed container
- black permanent marker
- plastic metric ruler – 30 cm
- pebble about one centimeter in diameter *(handwritten: (marble))*
- flashlight
- stopwatch

Part B
Materials Needed (per student group):
- two Slinkys® *(handwritten: large)*

Part C
Materials Needed (per student group):
- large open floor area (paved parking lot or playground)
- piece of white chalk (or masking tape)
- piece of red chalk *(handwritten: v. lg. protractor)*
- clear container
- long pencil
- Photocopy of **Blackline Master** *Our Dynamic Planet* 2.2, Modeling Wave Refraction
- (optional – clipboard)

Part D
Materials Needed (per student):
- pencil with a good eraser
- transparent straightedge or ruler
- Photocopy of **Blackline Master** *Our Dynamic Planet* 2.3, Refraction of Earthquake Waves

NOTES

INVESTIGATING OUR DYNAMIC PLANET

Investigation 2:

The Interior of the Earth

Explore
Questions

 Key Question

Before you begin, first think about this key question.

What is the interior of the Earth like?

Materials Needed

For this part of the investigation your group will need:

• large, flat-bottomed container

• black permanent marker

• plastic metric ruler – 30 cm

• pebble about one centimeter in diameter

• flashlight

• stopwatch

You now understand how sensory observations and experiments provide the basis for models. Make a drawing of what you think the interior of the Earth is like. What measurements, observations, or instruments would give scientists evidence about the Earth's interior?

Discuss your drawing and your thinking with your group and your class.

 Investigate

Part A: Observing Waves and Measuring Wave Speed

Design
Investigations

1. With the permanent black marker, make a clearly visible dot near one end of the bottom of a container. Make another clearly visible dot near the other end of the container. Measure the distance between the dots, in centimeters.

a) Record the distance in your journal.

Key Question

Use the **Key Question** as a brief warm-up activity that draws out students' ideas about the topic explored in **Investigation 2**. This question is designed to find out what your students know about the interior of the Earth and to set the stage for inquiry.

Write the **Key Question** on the blackboard or overhead transparency. Draw a large circle and ask students to do the same in their journals. Encourage students to think about what they know about the inside of the Earth, and what they learned about gathering evidence in the last investigation. Tell students to record their ideas in a new journal entry. Remind them to not look ahead into the investigation (where diagrams of the Earth's interior are shown), because you want to know what they currently think.

Discuss students' ideas. Ask for a volunteer to record responses on the blackboard or overhead projector. This allows you to circulate among the students, encouraging them to copy the notes in an organized way.

Student Conceptions about the Earth's Interior

Students' drawings will likely note that the interior of the Earth is hot and under high pressure. They will often note that there is magma inside the Earth. Some students will have learned that the Earth has a layered interior and will recite the three main layers—crust, mantle, and core. Few middle school students fully understand the variations in the properties of these three layers or comprehend the relative thickness of individual layers. Many students believe that the mantle is largely molten and soft, when in fact it is largely solid and quite rigid relative to liquids. In terms of thinking about the kinds of measurements that would give us evidence about the Earth's interior, students are most likely to think about drilling into the Earth, going into mines, or studying earthquakes.

Answer for the Teacher Only

The Earth's interior is known, from the study of how seismic waves travel through the Earth, to have a layered structure. The innermost part of the Earth, from a depth of about 5150 km to the center of the Earth, at about 6370 km, is called the core. It consists mostly of iron. The outer part is liquid, but the inner part is solid. The mantle, which extends from below the crust to the top of the core, consists of rock. Above the mantle is the crust, also consisting of rock, which ranges from a few tens of kilometers in thickness, beneath the continents, to several kilometers in thickness, beneath the ocean basins. Both temperature and pressure increase steadily downward in the Earth. The pressure increases downward because of the weight of the overlying material at any given level. The temperature increases downward because of interior sources of heat (both original heat, left over from the formation of the Earth, and radioactive heat, which is produced continuously by decay of natural radioactive materials in the Earth's interior).

Investigate

Teaching Suggestions and Sample Answers

Part A: Observing Waves and Measuring Wave Speed

Introducing Part A of the Investigation

You will need about 20 to 30 minutes to complete the "hands-on" part of the investigation (**Steps 1–6**). After students complete **Step 6**, they can return to their desks or put materials away and then complete the calculations. Doing so splits the investigation into a "wet lab" and "dry lab" and makes it less likely that water will be spilled onto papers.

Let students know that scientists use earthquake waves that pass through the Earth to make models of what the Earth is like. In some ways, this is like an x-ray of the Earth, because like earthquakes, an x-ray source sends energy into an object and the energy travels in the form of waves. In making a model of the Earth, it is important to find out the time it takes for an earthquake wave to travel from the location of the earthquake to a recorder (called a seismograph). How long the wave takes to travel, and whether or not the wave makes it to the seismograph, provide important evidence about the interior of the Earth. In this investigation, students will focus on measuring the speed of waves in water.

Review the main steps of the investigation. Students will be measuring the distance between two points, measuring how much time it takes for a wave to travel from one point to another, and calculating the average travel time and average speed of the waves. You might have them read through the steps of the investigation and produce the data tables necessary to record their measurements. This will help students to learn the value of planning ahead, and it may also lead to some questions about the activity, which you can clarify in a class discussion.

Students may not understand this concept at first. They need to become aware that for test results to be accurate, the test itself must be free of uncontrolled variables, other than those that are intentionally uncontrolled. In this instance, it is crucial that the test be conducted in exactly the same way every time if it is to be counted as reliable (or, in the language of scientific experimentation, "fair"). Students do have a strong sense of "fairness" in their daily lives, as in "taking a fair share", "being treated fairly", and so on. You can use these examples as an analogy in helping them to understand what a fair test is.

An important aspect of science that you might wish to review with students is when to "throw data out". Sometimes part of the experiment doesn't go according to plan. In this activity, making accurate measurements is important. If students think that they have not made a "fair and objective" measurement (e.g., the stopwatch was not started at the instant that the pebble hit the water), they should not use this data in their calculations. Students have to use their best judgment about when to do this, as do scientists. It's acceptable—and usually highly desirable— to do a few practice runs and get the procedure well worked out. Even then, there may be mistakes to factor in when doing calculations. Ask students to identify some of the possible sources of error (dropping the pebble from variable heights, looking at the wave from different angles during each trial, not starting and stopping the stopwatch on time, etc.).

Remind students about safety precautions. Pebbles should not be thrown, and safety goggles should be worn. Spills should be reported to the teacher and wiped up promptly.

1. A distance of at least 40 cm between points is recommended (somewhat farther than what appears in the diagram on page P9 of the student text). Students need to record the distance between the two points as accurately as possible. As you circulate, check measurements for accuracy. This is an excellent opportunity to help students refine their skills in measurement.

2. Pour water into the container to a depth of 2 to 3 cm. Let the water come to rest, until the water surface is mirror smooth.

3. Have one student hold a pebble 5 cm over one of the marks on the bottom of the container.

Conduct Investigations

4. Have a second student shine a flashlight beam straight down on the other mark. This student should lean directly over the flashlight to see the reflection of the beam from the water surface. This student should also hold a stopwatch and be familiar with how it works.

5. While the second student stares carefully at the reflection of the flashlight beam, the first student drops the pebble into the water. The second student starts the stopwatch when he or she hears the pebble enter the water. The second student stops the watch when he or she first detects motion of the water in the flashlight beam. This motion signals the arrival of the wave.

Inquiry

Conducting an Investigation

After scientists have designed an investigation, they conduct their test. The test must be free from uncontrolled variables. In this case you must be sure that the pebble is dropped from the same height each time, and the stopwatch is started and stopped at the correct time. Tests are often repeated several times to ensure reliable and valid results.

Wear goggles when dropping the pebble. Wash your hands after the activity.

Collect & Review

a) Record the time on the stopwatch as the travel time of the wave.

2. Water need not be very deep in the containers. Two to three centimeters (about one inch) works just fine.

3. As you circulate, check for consistency of the height from which students are dropping the pebble. Five centimeters is about two inches.

4. Make sure that each student takes an active role in the investigation. In addition to the two students doing the measuring, a third can be checking the quality of the trial (was the stopwatch used in a timely way, was the pebble dropped from the same height on successive trials?). A fourth can record data, noting any questionable measurements.

5. Circulate around the room and spot check the quality of students' work. Is data being recorded clearly? Do students know how to read the stopwatches? Are trials being run properly? Are students recording units (seconds) as well as values?

   a) Students also need to know that the accurate recording of information is crucial in science. Help them to see that using a data table is a good way of organizing this recording. Students also need to understand that the method of recording has to suit the test being conducted and that there are a variety of "tools" for doing this, including drawings, charts, graphs, and, in some instances, specialized computer software. A sample student data sheet is provided below.

Sample student data sheet

October 28
Data Table – Wave speed investigation (water waves)
Group Members: Pam L., Jen M., Brooke B., Terry W.

Trial Number	Distance Wave Traveled (centimeters)	Wave Travel Time (seconds)	Notes about Quality of Measurement
1	45.5	1.37	Good
2	45.5	1.31	Good
3	45.5	1.32	Good
4	45.5	1.40	Good
5	45.5	1.69	Questionable – starting the stopwatch before pebble hit the water.
6	45.5	1.35	Good
7	45.5	1.29	Good
8	45.5	1.41	Good
9	45.5	1.34	Good
10	45.5	1.19	Questionable – stopped timing too soon.

INVESTIGATING OUR DYNAMIC PLANET

6. Repeat the measurements until your group has at least 10 measurements. If one or two of the measurements are very different from all of the others, they are probably what a scientist would call a "gross error." You can ignore a few bad measurements like that, but be sure you have a large number of "good" measurements.

7. Calculate the average travel time. Add up all the measured travel times and then divide the sum by the number of measurements. Be sure to not include the bad measurements.

a) In your journal, record the average travel time.

8. Calculate the wave speed in centimeters per second. Use the following equation:

$$\text{average wave speed (cm/s)} = \frac{\text{distance between dots (cm)}}{\text{average travel time (s)}}$$

a) Record your calculations and your result in your journal.

9. Suppose you had a long pan of water. How long would it take the waves to travel:

a) 50 cm?

b) 100 cm?

c) 200 cm?

10. Suppose you dropped stones into a material through which waves move twice as fast as they do through water.

a) How would this change the average travel time of the waves?

11. Scientists cannot observe earthquake waves moving through the Earth in the same way you can observe waves moving through water. They can, however, record and study the energy from earthquake waves as the waves arrive at a recording station (seismograph station). They can use information they record about the waves to make models of the interior of the Earth.

Think about how what you studied relates to how scientists make models of the inside of the Earth. What part of your experiment represented:

a) An earthquake, which releases energy in the Earth?

b) The movement of energy waves from the earthquakes (seismic waves) in the Earth?

Inquiry

Using Mathematics

Scientists use mathematics in their investigations. Accurate measurement, with suitable units, is very important when collecting data. In this investigation you made measurements. You then used the measurements to make calculations to interpret the data you collected.

Use Mathematics

Use Mathematics

Evidence for Ideas

Evidence for Ideas

6. Keep in mind that the water waves are small and fast. Stress that running the test several times and averaging the results is a common practice in scientific tests to ✓ ensure accuracy. An important aspect of science that you might wish to review with students is when to "throw data out." Sometimes part of the experiment doesn't go according to plan. Remind students to record the criteria that they used to make decisions about "good measurements" versus "gross error."

7. In making their calculations, students should not include bad measurements. For example, in the sample student data table, trial 5 was too high and trial ✓ 10 was too low.

   a) Answers will vary according to the actual distance between the points. The average travel time for the eight "good measurements" in the sample data table is 10.79 s divided by eight trials, or an average of 1.35 s per trial.

8. Circulate as students make their calculations. Sometimes students rely on one group member to "do the math." Encourage all students to be involved, and to "double check" the accuracy of their work within their groups.

   a) Answers will vary. For the sample data table, the average wave speed is 45.5 cm divided by 1.35 s, or 33.7 cm/s.

Teaching Tip

Students here are dealing with speed, or travel distance over time, which requires calculating an average from a set of data. These can be difficult mathematical concepts for some students to master. Check with your student's mathematics teacher(s) to see if any coordination is possible.

9. Here students must use the average wave speed and make predictions about the arrival times for various distances. Scientists use their understanding of the average speed of seismic waves to determine how far away an earthquake is from a seismograph station. If average wave speed = distance divided by average travel time, then average travel time = distance divided by average wave speed. Middle school students working in groups should be able to reason this through. Allow students some time to discuss this challenge. Provide assistance if necessary.

Results for the sample data table are provided below.

   a) 50 cm divided by 33.7 cm/s = 1.48 s

   b) 100 cm divided by 33.7 cm/s = 2.97 s

   c) 200 cm divided by 33.7 cm/s = 5.94 s

10. This question prompts students to think about the speed of waves in a different material. If students understand the meaning of average speed, then they should be able to reason through this question.

   a) If the speed of the waves is twice as great, then it would take half the time to travel the same distance.

11. This question helps students to begin to connect their understanding of wave energy and wave speed to using earthquake waves to study the Earth's interior.

   a) Dropping the pebble released energy into the water to form water waves, in much the same way that an earthquake releases energy in the form of earthquake waves.

   b) The movement of the water waves away from the splash of the pebble was like the movement of earthquake waves away from the earthquake.

NOTES

Investigation 2: The Interior of the Earth

c) The material in the Earth through which seismic waves travel?

d) The arrival of a seismic wave at a seismograph station where earthquakes are detected?

Seek Alternatives

12. Compare your average travel time with those of other groups. Discuss the following questions and record the results of your discussions:

Show Evidence

a) What might cause differences in travel times from *apparent* measurement to measurement within your group?

b) What might cause differences in average travel times among the different groups?

c) What improvements to your measurement technique might decrease the difference in values you obtained?

Part B: Kinds of Seismic Waves

Conduct Investigations

1. With a partner, stretch out a Slinky® on the floor about as far as it can go without making a permanent bend in the metal.

2. Have one partner make waves by holding the end of the Slinky with a fist, and striking the fist with the other hand, directly toward the end of the Slinky.

Collect & Review

Observe the direction of wave movement, relative to the Slinky.

a) Does it move in the same direction (parallel to the Slinky) or in the opposite direction (perpendicular to the Slinky)? Record your observation in your journal.

Evidence for Ideas

b) This kind of wave is called a compressional wave. (To compress means to squeeze together.) From your observations, explain why this is an appropriate name for this wave. You may wish to use diagrams to illustrate your answer.

Materials Needed

For this part of the investigation your group will need:

• two Slinkys®

Wear safety goggles throughout. Be sure that neither partner holding the Slinky lets go while it is stretched out. Wash your hands after the activity.

c) The water represented the material in the Earth.

d) The arrival of the water wave at the flashlight beam was like the arrival of the earthquake wave at a seismograph.

12. Think about having students share their findings. Make an overhead transparency of average travel times and average wave speeds for each student group (an example is provided below). The data table below is provided as **Blackline Master** 2.1, Class Data. This is a great opportunity to discuss subjectivity and variability in scientific measurement, an important aspect of the nature of science.

Class Data – Measuring Speed of Water Waves

Group Number	Average Travel Time (seconds)	Average Wave Speed (centimeters per second)
1		
2		
3		
4		
5		
Class Average		

a) Differences in travel time from measurement to measurement can come from the accuracy with which the stopwatch is started and stopped.

b) Differences in average travel times between groups are probably due to the differences between the ways that different students start and stop the stopwatch. Different people have different "reaction times" to events, or they may interpret the arrival of the wave at the flashlight differently.

c) Answers may vary. One possible response is to increase the distance between the points so that there is more time between the start and stop of the stopwatch, which should make it easier to record the arrival time. Another way would be to have groups work with one another to discuss differences in technique, in order to arrive at a consensus about the very best way to make the measurements.

*use of
diagrams*

Part B: Kinds of Seismic Waves
Introducing Part B of the Investigation
In **Part A** of the investigation, students studied one kind of wave—a wave that moves
in water. In **Part B**, students use Slinkys® to observe two kinds of seismic waves:
compressional waves and shear waves. Students will see that compressional waves
travel differently than shear waves, as well as travel faster than shear waves. Scientists
study the arrival times and amplitudes (height of the waves from trough to crest) of
seismic waves to study the interior of the Earth.

Show students a Slinky. Ask them to think about the Slinky being stretched out as
far as it can go without damaging it, and to tell you how many different ways that
wave energy can be sent from one end of the Slinky to the other. Students will likely
respond that you can hit one end of the Slinky, and you can shake the Slinky side to
side (back and forth), and you can shake the Slinky up and down. Remind them
about the measurement of wave speed that they made in **Part A**. Tell students that
they are going to study the movement of waves through the Slinkys to see if there are
any differences in how the waves move and how fast the waves travel. **Part B** of the
investigation will require about ten minutes for students to complete.

1. The long Slinkys can be stretched about ten meters.

2. When the Slinky is stretched out nearly as far as possible, one student should hold
   the end of the Slinky in his or her fist and then strike the fist in a direction moving
   toward the end of the Slinky (i.e., toward the person who is holding the other
   end). This will send a compressional wave down the Slinky and back. An
   alternative is to bunch up about a dozen coils of the Slinky and then quickly let
   them uncoil.

   key ⟶ a) The wave moves parallel to the Slinky.

   b) The wave is called a compressional wave because some of the coils get
   dilate compressed, then they decompress (expand or "uncoil"), sending energy down
   the Slinky, which gets compressed, uncoils, and repeats until the end of the
   Slinky is reached.

Teaching Tip

Circulate from group to group, asking questions that help the students better understand the process of vibration. Some guiding questions might be:

• Did you push–pull or shake the Slinky?

• What determines whether the wave travels the length of the Slinky? (If the Slinky is not hit hard enough, friction causes the wave to die out before it reaches the other end.)

• How do the coils change as the wave passes? (They are compressed together as the push–pull wave passes. They slide past each other sideways — that is, they are sheared — as the shake wave passes.)

• What happens when the wave reaches the other end? (It bounces back [it's reflected] off the other person's fist.)

• How does the wave travel compared to the orientation of the Slinky? (Push–pull waves travel parallel to the direction of the Slinky. Shake waves travel perpendicular to the orientation of the Slinky.)

• Do you think waves travel through rocks? (Yes. Think about a passing truck, or about striking the ground with a sledgehammer. A person standing nearby would feel vibrations.)

INVESTIGATING OUR DYNAMIC PLANET

3. Stretch out the Slinky again. This time, have one partner make waves by moving the Slinky from side to side (left to right or right to left).

Again, observe the direction of wave movement, relative to the Slinky.

a) Does the wave move in the same direction (parallel to the Slinky) or in the opposite direction (perpendicular to the Slinky)? Record your observation in your journal.

b) This kind of wave is called a shear wave. (To shear means to slide one thing sideways past another thing.) From your observations, explain why this is an appropriate name for this wave. You may wish to use diagrams to illustrate your answer.

4. To compare the two types of wave motions, stretch out two Slinkys along the floor, about 5 m.

Starting at the same end at the same time, have one student holding the end of the Slinky strike their fist. Have another student jerk the other Slinky back and forth. Observe what happens.

Try the movements several times until you are confident in your observations.

a) Which of the two wave types arrives at the other end first (which one is faster)? Explain why you think this happens.

A stretched Slinky can move unpredictably when released. Spread out so that you can work without hitting anyone. Do not release the Slinky while it is stretched. Walk together until the Slinky is not stretched and then have one person take both ends to put it away.

Materials Needed

For this part of the investigation your group will need:

- large open floor area (paved parking lot or playground)
- piece of white chalk (or masking tape)
- piece of red chalk
- clear container
- long pencil

Part C: Refraction of Waves

In **Part A**, you investigated the speed of waves as they pass through one kind of material (water). In **Part B**, you saw that different kinds of waves have different speeds. Next, you will simulate what happens when a seismic wave crosses a boundary between two kinds of materials. It will help if you read all the steps below before you do the activity.

1. One person will be the "marker." That person holds the pieces of chalk. The marker should draw two lines:

- A long, straight line in white chalk across the middle of the area. The white line represents a boundary between two layers in the Earth.
- A long straight line in red chalk at a 20° angle to the white chalk line. The red line represents the front of a seismic wave that is moving through the Earth.

Collect & Review

Evidence for Ideas

Evidence for Ideas

Conduct Investigation

P
12

3. The diagram on page P11 shows students making a shear wave in the Slinky. It's important that one person hold the Slinky steady, and the other generate the waves. If both students are generating waves, it will be harder to evaluate the wave motion.

    shear — waves move ⊥ to Slinky

    a) The wave moves in a direction perpendicular to the Slinky.

    b) Students may be confused by the meaning of the word shear. You can tell them that shear means motion perpendicular relative to the direction that the wave is moving. The Slinky must twist and distort itself slightly to accommodate the movement of the wave.

    key

4. This will take some practice. The trick is for one student to try to impart an equal amount of energy into both Slinkys (students who shake the Slinky back and forth can get carried away at times). Have one student act as the impartial observer. Encourage students to do multiple trials until the observer is sure about the result. A second observer should then observe and see if he or she agrees with the first.

    a) The compressional wave arrives before (travels faster than) the shear wave.

Part C: Refraction of Waves

Introducing Part C of the Investigation

This investigation should take less than one class period to complete. If you give students an overview of the investigation before they get started on the activity, things will go much more smoothly. This is especially important if you are going to take them outside to do the activity.

Distribute a copy of **Blackline Master** *Our Dynamic Planet* 2.2, Modeling Wave Refraction to each student. Ask students to read through **Steps 1** through **8**. Point out the diagram on page P13. Tell students that the diagram shows a row of students who will march toward a "boundary line." The row of students represents the front of a wave that is moving through the Earth. The white line represents a boundary between two kinds of materials. As each part of the wave crosses the boundary line, its speed changes, because the speed of a wave depends upon the material through which it is traveling. This holds for compressional waves as well as shear waves. The question for students to answer in this activity is: Does a change in the speed of a wave have any effect on its path or direction? When an earthquake happens, if the seismic waves move through one kind of material in the Earth and the material has a constant density with depth, we would expect the waves to have a constant speed (like the water waves in **Part A**) and follow a straight path.

Ask students to predict what would happen if they did not change the speed of their marching after they crossed the white boundary line. This serves as a thought experiment that would provide an experimental control — what happens when none of the variables are changed. They should respond that the angle of the second red line and white line should be the same. Then ask students to think about what would happen if they increased the speed of their marching after crossing the boundary line.

Have them predict what the line of students will look like after the last person has crossed the boundary. Have them draw and explain their prediction on a copy of **Blackline Master** *Our Dynamic Planet* 2.2, Modeling Wave Refraction. Hold a brief discussion about their ideas. Then go to where you will do the activity. If you do the activity outdoors, students should have a notebook or clipboard to provide a solid writing surface for their work.

1. Have a student draw the long straight line. The length of the line will depend upon how many students you want in one group. The more students in one group, the more challenging it is to keep them all walking properly. For the best results, do not exceed eight students in one group. For a group of eight students, the white line will need to be at least 12 m long (1.5 m per student).

NOTES

The marker should also draw an arrow about one meter long and perpendicular to the red chalk line. The result should look like the diagram shown.

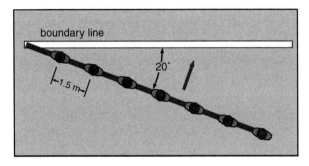

2. Form a line of students along the red line. Stand about 1.5 m apart from one another (arm's length apart).

3. When the marker gives the signal to start, everybody in the line moves forward, taking steps 30 cm long every second. Walk straight ahead as shown by the arrow in the diagram.

4. The instant you cross the white line, keep walking straight ahead, but start to take steps one meter long instead of small steps.

5. After the last person has reached the white chalk line, the marker will tell everybody to stop walking.

6. With the red chalk, the marker will make a long, straight chalk line just in front of everyone's toes. The new red line should connect to the white chalk line. The marker should also draw an arrow about three feet long and perpendicular to the new red line.

7. Stand back and look at the two red chalk lines. Compare the angle of the new red chalk line with that of the old red chalk line.

8. Obtain a copy of the diagram shown above.

   a) Label the region above the boundary "wave speed = 1 m/s" and the layer below the boundary "wave speed = 30 cm/s."

   b) The diagram already shows the old red chalk line. Add the new red line from your results to the diagram. Try to draw the angle accurately, but do not worry if it is not exact.

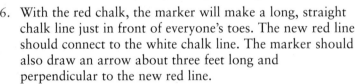

The arrow should be about one meter long. It is used to indicate the direction in which students should walk (straight ahead). It also indicates the original direction of the wave, which will be useful for comparing the change in the direction of the wave front after all students have crossed the line and another arrow is drawn.

2. Make small marks on the red line about 1.5 m apart — these are markers at which students should stand so that they are evenly spaced.

3. Having the marker count off each second (one, two, three…) will help to pace students. You might want to have a student demonstrate what a 30 cm step looks like compared to a meter step, or demonstrate this for students yourself. You might also demonstrate how the length of the step changes, but not the direction of the marching, after crossing the white line.

4. Observe students to make sure that they follow directions. If someone makes a mistake (forgets to take large steps after crossing the line), have students start over.

5. Students should stop walking after the marker has called out "Stop!" Tell students to remain in place.

6. While students remain in place, the marker draws a red chalk line as straight as possible along the front of everyone's feet and extends the line until it touches the white line. The marker should also draw an arrow perpendicular to this new red line.

7. Have a student measure the angle between the new red line and the white line. Students will see that the direction of the new arrow has shifted to the right, and that the angle between the new red line and the white line is greater than the angle of the original red line.

8. Students should record what they have observed on a copy of **Blackline Master** *Our Dynamic Planet* 2.2, Modeling Wave Refraction.

    a) **and** b) A sample student paper is shown below:

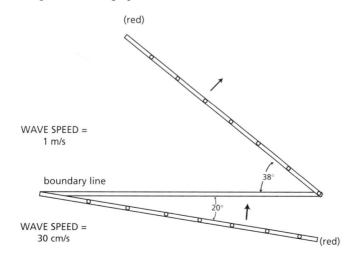

INVESTIGATING OUR DYNAMIC PLANET

Use Mathematics 5+3

c) Does the second red chalk line form a different angle with the white chalk line, or the same angle?

Evidence for Ideas

d) What is the basic reason for the difference in the angles of the red chalk lines?

e) If the speed of a seismic wave increased after it crossed a boundary in the Earth, what do you think would happen to the wave?

Explore Questions

9. With your group, devise a plan to investigate a different kind of change in wave speed. For example, in the original activity, the waves moved (roughly) 30 cm/s and then moved 1 m/s after they crossed the boundary. What would happen if the waves slowed down after crossing the boundary? What would happen if the waves moved four times as fast after crossing the boundary? How would this affect the path of the wave?

Design Investigations

a) In your journal, record what you plan to investigate.

b) Predict how you think this would this change the angle of the second red line. Include the reason for your prediction.

Have your plan checked by your teacher before you begin.

c) Record your prediction on a second copy of the diagram. Label the diagram to indicate the relative speeds of waves before and after the boundary.

Conduct Investigations

10. Repeat the activity.

a) Record your observations in your journal by drawing the angle of the new red line.

b) How does the angle of the new red line compare with your prediction?

Evidence for Ideas

Clean up spills immediately.

11. Fill a clear container almost full of water. Put the pencil halfway into the water, at an angle. Look at the pencil from above, but slightly to the side.

a) Draw a picture of what you observe.

b) Do you think that light waves travel at the same speed through water as they do through air, or at different speeds? Explain your answer. Relate this to what you discovered about the bending of waves in **Part C.**

Evidence for Ideas

12. Imagine a liquid in which light waves travel twice as fast as they do through water.

a) Draw a picture of what the pencil would look like if it were put into this liquid. Explain your drawing.

P
14

Investigating Earth Systems

c) The second red chalk line forms a different angle with the white chalk line than the first red chalk line does.

d) The change in speed of the wave after crossing the boundary makes the overall direction of the wave change.

e) If the speed of a seismic wave increases as it crosses a boundary line in the Earth, the direction in which the wave is moving changes. This means that the wave bends.

9. Encourage students to be creative. Samples of student plans include changing from a fast wave speed (1 m/s) to slow wave speed (30 cm/s); starting at a very slow wave speed (10 cm/s) to a very fast wave speed (1.5 m/s). To some extent, the variations that students can accomplish will be limited by the smallest "longest stride" of any member of their group.

a) Answers will vary. Sample plan:
We want to know how the path of the wave changes if the initial speed of the simulated wave is 100 cm/s and the speed of the wave after it crosses the boundary is 50 cm/s (a slowing of the wave velocity by one-half). We will take steps 100 cm long before we reach the boundary line and steps 50 cm long after we cross the boundary line. We will measure the angle of the wave compared to the boundary line and compare it to our original investigation.

b) Sample hypothesis: We think that the angle of the second red line will change, and that it will be lower than the original angle (less than 2°). The angle will be smaller because the left end of the wave will be moving slowly once it crosses the boundary, and the right end of the wave will be in the process of "catching up" with the left end.

c) Answers will vary. Work should be labeled "prediction" with the values for the initial speed and second speed, the names of the lines, the original angle, the predicted angle, and a second arrow to indicate the direction of the wave. A sample is provided below.

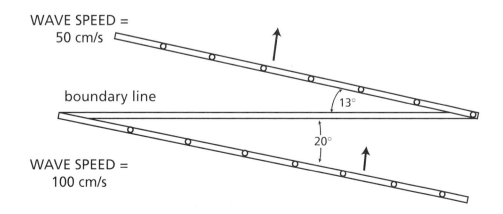

10. Check students' work before allowing them to proceed with the activity.

   a) Students should measure the angle of the wave and record the information in their journal. They can include this right on their drawings of the predicted angle. Have them use a dashed line to indicate that this is the actual result compared to the predicted result.

   b) Answers will vary depending upon the investigation students are doing. Answers should include a comparison.

Assessment Tool

Investigation Journal Entry-Evaluation Sheet

Use this sheet to help students learn the basic expectations for journal entries that feature the write-up of investigations. It provides a variety of criteria that both you and your students can use to ensure that their work meets the highest possible standards and expectations. Adapt this sheet so that it is appropriate for your classroom, or modify the sheet to suit a particular investigation.

Teaching Tip

The first part of **Part C (Steps 1–10)** simulated refraction. **Step 11** shows refraction. In this case, a pencil inserted into a container of water will appear to bend because light waves travel at a different speed through air than through water. This bending is the result of refraction.

It would be a good idea for students to share the results of their investigations from **Steps 1–10** in a class discussion. The discussion will help them search for patterns and relationships among the different investigations that groups conducted.

In general, when wave speed decreases after crossing a boundary, the wave is refracted in such a way that the wave front makes a smaller angle with the boundary (in other words, it becomes more nearly parallel to the boundary), and when the wave speed increases after crossing a boundary, the wave is refracted in such a way that the wave front makes a larger angle with the boundary. Things work this way regardless of the original angle of approach, and also of the original direction of approach (that is, whether the angle between the wave front and the boundary "opens" to the left or to the right).

11. Students should observe the pencil or chopstick from above and slightly to the side.

   a) Students should make a drawing of what they observe (the change in angle of the stick as it goes into the water).

b) Students should be able to reason that this demonstration is related to their simulations of changes in wave speeds on the playground. The surface of the water is a boundary between air and water. In order for them to see the stick, light waves must travel first through air *water* and then, after being refracted, through water *air*. If you did the suggested thought experiment outlined in this guide (see **Introducing Part C of Investigation 2**), students will know that the angle would not change if light waves travel at the same speed through air as they do through water.

12. This is a thought experiment that checks whether or not students understand the general relationships between changes in wave speeds and refraction. When a wave crosses a boundary between two different materials, the greater the change in speed the greater the refraction.

    a) Students' drawings should note a greater angle of the stick in the liquid than what they observed when the stick was in water. A sample is provided near the answer to **Step 11(b)** above.

Part D: Refraction of Earthquake Waves in the Earth

1. Obtain a copy of a diagram similar to the one shown.

    The large circle represents the Earth. The small circle represents the edge of an inner part of the Earth where earthquake waves move ~~faster~~ slower than in the outer part. The black dot near the right-hand edge of the large circle represents a place in the Earth where an earthquake happens.

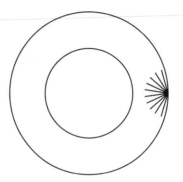

Materials Needed

For this part of the investigation your group will need:

• pencil with a good eraser

• transparent straightedge or ruler

2. The earthquake sends seismic waves in all directions through the Earth. Notice the short lines coming from the dot. These lines show the start of some of the directions in which the waves start to travel (like the arrow that you drew in **Part C**). Your challenge is to show how these directions change when the wave goes through different layers of the Earth.

3. Use your straightedge and pencil to extend the lines through the Earth to the other side. Think about the following before you begin:

    • Some of the lines will go through the Earth without hitting the inner circle.

    • Some of the lines, however, will hit the inner circle. This is a boundary between zones with different wave speeds. Assume that the speed of the waves decreases as the waves cross this boundary. From **Part C** of the investigation, when you pretended you were part of a wave, you then learned what happens.

    • The lines that go into the inner circle also come out of the inner circle. When that happens, the waves will be crossing a boundary again. Think about what will happen to their direction of travel.

4. When you have completed all of the lines, look at where they are when they reach the other side of the Earth.

    a) Describe the pattern shown by the waves when they reach the other side of the Earth.

Part D: Refraction of Earthquake Waves in the Earth

Introducing Part D of the Investigation

This part of the investigation should require about twenty minutes to complete. Distribute a copy of **Blackline Master** *Our Dynamic Planet* 2.3, Refraction of Earthquake Waves and materials to each student. Tell students that they will need to use what they learned about wave refraction in **Part C** of the investigation to predict what would happen to the paths of earthquake waves as the waves pass through the Earth. Tell them that for the purpose of this investigation the earthquake waves are compressional waves (waves that can travel through solids, liquids, and gases). Remind students about how this is also an example of how models can be used to make predictions (**Investigation 1**). They have investigated models of what happens when waves cross a boundary between two kinds of material, and they are now being asked to use their models to predict what would happen to earthquake waves if the waves pass through two kinds of materials in the Earth. (Note: the drawings that students produce will resemble the general idea shown on the diagram of seismic waves shown on page P19 of the student text.)

IMPORTANT NOTE: Tell students that *the speed of the seismic waves decreases as the waves cross the boundary represented by the inner circle*. Otherwise, the exercise is ambiguous.

Students should be able to do this activity on their own with some assistance from you. Circulate to groups as students complete the activity. Suggestions for helping groups to understand steps of the activity are described below.

1. Check with groups to see if they have any difficulty understanding what the diagram represents.

2. Point out to students that the lines on the right side of the circle represent the directions in which earthquake waves move. Seismic waves radiate out in all directions from an earthquake. Lines drawn perpendicular to this expanding front show the direction of movement of the waves and look like spokes on a wheel when viewed in two dimensions.

3. Check student work. Lines should be refracted when they enter the inner circle and be refracted again when leaving the inner circle.

4. a) Students should note that there is a gap on the other side of the Earth where no waves reach the Earth's surface. A sample diagram has been provided on the following page.

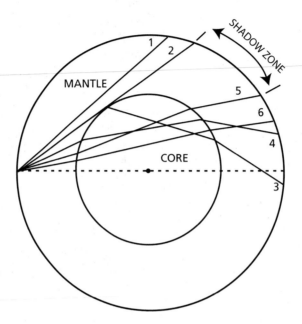

NOTES

b) How does the pattern help you to understand how scientists in another part of the Earth can detect an earthquake?

Evidence for Ideas

c) How does this help you to understand why scientists in some places would not be able to detect some of the earthquake waves?

d) Scientists used this pattern to argue that the Earth has a core. They claimed that it showed that seismic waves pass through a zone with different wave speeds. Does this conclusion make sense? Explain why or why not.

As You Read...
Think about:

1. *What is the difference between compressional (P) waves and shear (S) waves?*

2. *How do earthquakes produce seismic waves?*

3. *How would you describe wave refraction?*

4. *What is the focus of an earthquake?*

5. *How are earthquake waves detected on the surface of the Earth?*

6. *How do scientists know that the Earth's mantle is made of solid rock?*

7. *How do scientists know that the Earth has a core?*

Digging Deeper

Evidence for Ideas

WHAT EARTHQUAKE WAVES REVEAL ABOUT THE INTERIOR OF THE EARTH

Waves

Shaking a material produces vibrations. Those vibrations move away in all directions in the form of a wave. You saw that clearly with the Slinky®. Speech is another good example. When you speak, you are making your vocal cords vibrate. That makes the air around them vibrate. The vibrations travel out from your mouth and through the air as sound waves. As the waves travel, they carry energy with them. The water waves you made in the pan were given their energy by the falling stone. The waves then delivered their energy to the sides of the pan.

The first kind of wave you made with the Slinky is called a compressional wave. When you compressed (squeezed together) the end of the Slinky by hitting it with your fist, it tried to expand again, and when it did, it compressed the material next to it. That part of the coil then expanded again, and so on. The wave of compression and expansion traveled along the coil as a wave.

b) The seismic waves travel through the Earth, are refracted, and reach the Earth's surface in another part of the world. The drawing reveals that seismic energy released within the Earth eventually resurfaces in many locations throughout the Earth, not just on the surface directly above the source.

c) If all of the seismic waves took straight paths through the Earth, then they would be detected all over the Earth. Because the waves are refracted, there are areas on the other side of the world that do not receive the seismic waves.

d) Refraction of waves occurs when a wave passes from one medium to another in which the wave travels at a different speed. If seismic waves are refracted inside the Earth, then there must be a core of material where seismic waves change speed.

Digging Deeper

This section provides text and illustrations that give students greater insight into the nature of waves, earthquakes and seismic waves, wave refraction, and the Earth's interior. You may wish to assign the **As You Read** questions as homework to help students focus on the major ideas in the text.

As You Read...

Think about:

1. Compressional waves move by compressing and expanding the material through which the wave travels. The compression and expansion happen in a direction parallel to the direction in which the wave travels. Shear waves move by pulling the material back and forth in a direction perpendicular to the direction in which the wave travels

2. An earthquake produces powerful vibrations in the Earth. These vibrations move out in all directions from the Earthquake as seismic waves.

3. Refraction is the change in direction of a wave that happens when a wave changes speed when it crosses a boundary between two different materials.

4. The focus of an earthquake is the place in the Earth where the earthquake occurs.

5. Earthquake waves are detected on the surface by recording instruments called seismographs.

6. Scientists know that the Earth's mantle is made of solid rock because shear waves can be detected as passing through it. If the mantle were liquid, shear waves would not be able to pass through the mantle.

7. Scientists know that the Earth has a core because there is a zone, on the other side of the Earth from where an earthquake happens, where no waves are detected. The only way to account for that zone, which is called the shadow zone, is that there is a boundary deep within the Earth where the speed of seismic waves suddenly decreases. The body of material below that boundary is called the core.

Assessment Opportunity

You may wish to rephrase selected questions from the **As You Read** section into multiple choice or "true/false" format to use as a quiz. Use this quiz to assess student understanding and as a motivational tool to ensure that students complete the reading assignment and comprehend the main ideas.

NOTES

The second kind of wave you made with the Slinky is called a shear wave. When you moved the Slinky sideways, it pulled the material next to it sideways as well. In turn, that material pulled the next part of the Slinky sideways, and so on down the length of the Slinky.

Earthquakes and Seismic Waves

During an earthquake, large masses of rock slide past each other, making powerful vibrations. That also happens in human-made explosions. The vibrations move away in all directions through the Earth in the form of waves, called seismic waves. The seismic waves travel all the way through the Earth. When they reach the Earth's surface again, they can be detected with special instruments called seismographs. There are hundreds of seismograph stations all around the Earth.

The Richter Scale is a way of measuring the magnitude of an earthquake using seismograph records. Each whole number increase on the Richter Scale is an increase of 10 times in the size of the vibrations recorded and an increase of 31 times the amount of energy released! To give you some idea of the different magnitudes, anything less than a magnitude 2.5 is too small to be felt by humans. A magnitude 4.5 or over is capable of causing damage near the earthquake, and anything over a magnitude 7 is considered a major earthquake that is potentially very destructive.

Seismic waves weaken as they move through the Earth. That is why the distance from the earthquake must also be considered when calculating the Richter magnitude. There are two reasons that the seismic waves weaken. They spread their energy over a larger and larger area, just like the ripples in the pan of water. Also, they lose energy because they create some friction as they move through the Earth.

NOTES

INVESTIGATING OUR DYNAMIC PLANET

Wave Refraction

You saw in **Part C** of the investigation how a wave (the line of students) changes its direction when it suddenly passes from a material with low wave speed ("short steps") to a material with high wave speed ("long steps"). The change in direction is called refraction. Refraction also happens where the wave speed changes gradually rather than suddenly. Suppose you were in a marching band, and the leader yells, "Members on the right take giant steps, members on the left take small steps." You can imagine what would happen. The whole line would gradually curve more and more to the left. The same kind of thing happens with seismic waves in the Earth. The speed of seismic waves generally increases downward in the mantle, so the path of a wave curves upward as the wave passes through the mantle (see the diagram).

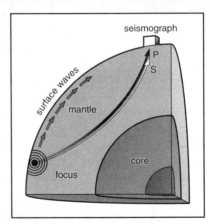

The reason why the speed of seismic waves increases downward in the mantle is complicated, but scientists are able to measure the seismic-wave speed of rocks in the laboratory using special equipment. Using this equipment they can determine how differences in temperature, pressure, and the type of rock affect the speed of seismic waves.

Show poster

About the Illustration

This schematic drawing shows the generalized paths of a compressional wave (P wave), a shear wave (S wave), and surface waves from the focus of an earthquake. The expanding rings around the focus of the earthquake represent how the wave expands outward in three dimensions from the focus (shown here as circles in this two-dimensional diagram). The P wave is shown to arrive first at the seismograph (P waves travel faster than S waves). The drawing also shows that surface waves do not travel as far (they die out more quickly). Also note that both the P wave and S wave are curved as a result of the refraction that occurs as the waves encounter rocks with higher seismic wave speeds with depth in the Earth's mantle.

sketch on board & in notebook

p319-321

Teaching Tip

You may wish to use the following **Blackline Masters** to make overheads to use when discussing the **Digging Deeper** reading section.

- *Our Dynamic Planet* 2.4, Seismic Wave Refraction and the Earth's Interior Structure
- *Our Dynamic Planet* 2.5, P wave Shadow Zone
- *Our Dynamic Planet* 2.6, Earth's Interior Structure

The Interior of the Earth

The main way that scientists know about the interior of the Earth is by studying how seismic waves pass through it. In **Part D** of the investigation, you showed something important about the interior of the Earth. Long ago, scientists noticed that there is a zone where seismic waves from a faraway earthquake do not appear again at the surface. This zone is in the form of a large ring on the Earth's surface, as shown in the diagram. This ring is called

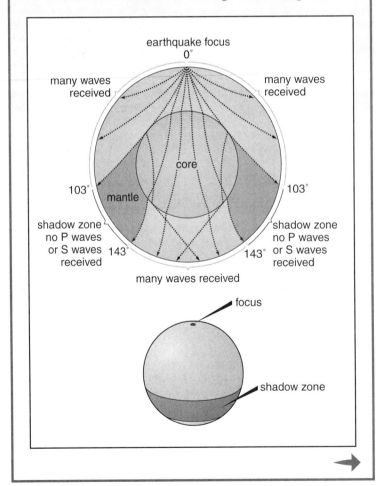

About the Illustration

This diagram shows the origin of the shadow zone. Waves that start out in a direction that makes a shallow angle with the Earth's surface pass through the Earth without encountering the core–mantle boundary. Waves that go down more steeply are refracted at the boundary. There is a ring-shaped zone (called the shadow zone) on the far side of the Earth where none of the waves emerge. Use this diagram as a guide when you are working with students on **Part D** of the investigation.

As you discuss schematic diagrams like the one shown on this page with students, keep in mind that students sometimes draw surprising conclusions from such diagrams. For example, field tests of this module revealed that some students concluded that earthquakes happen only at the top of the Earth!

the "shadow zone." This shows that there is a place deep inside of the Earth where there is an abrupt slowing of seismic-wave speeds. How can this be if the seismic-wave speed of the mantle increases with depth? The answer is that there must be an inner shell of the Earth (called the core) that is made of a material that is very different from the mantle.

The core almost certainly consists mostly of iron. You probably know, from using a compass, that the Earth has a strong magnetic field. The only way the Earth can have a magnetic field is for the core to be made of iron. The outer part of the Earth, above the core, is called the mantle. It consists of mostly solid rock. Remember from **Part B** of the investigation that two kinds of seismic waves can go through the Earth: compressional (P) waves, and shear (S) waves. It's known that P waves can pass through solids, liquids, and gases, but S waves can go only through solids. Both P waves and S waves go through the mantle, so it must be solid. On the other hand, only P waves go through the core, so its iron must be melted rather than solid. There is even an inner core within the outer core. The inner core seems to be solid rather than liquid. Scientists discovered the inner core by some very careful detective work, long after the core itself had been discovered. The outermost layer of the Earth, above the mantle, is called the crust. The crust is very thin, not more than about 50 km thick. You will learn more about the Earth's crust in a later activity. The diagram shows the Earth's core, mantle, and crust.

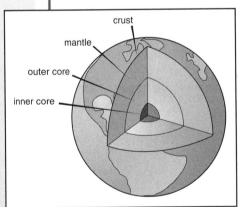

About the Illustration

This is a cutaway view of the overall structure of the Earth's interior. On the scale of this diagram, the crust is so thin that it is represented to be no thicker than a thin ink line.

Investigation 2: The Interior of the Earth

Review and Reflect

Review

1. The direction of wave motion and the direction that a wave travels or propagates are not necessarily the same thing. How were the two types of waves you made in **Part B** of the investigation different from one another?

2. Why do seismic waves follow a curved path through the Earth?

3. How do scientists know that the Earth's core is made of a different material than that of the mantle?

Reflect

4. Think about the **Key Question:** "What is the interior of the Earth like?" Has your answer to this question changed from the beginning of the investigation? Explain.

5. Give an example of how you used a model in this investigation.

6. What have you learned about how scientists investigate the Earth?

Thinking about the Earth System

7. On your *Earth System Connection* sheet, summarize what you have learned about the geosphere.

8. Suppose an earthquake occurred below the ocean floor. How might earthquakes affect the hydrosphere?

Thinking about Scientific Inquiry

9. How did you revise your ideas on the basis of evidence?

10. Why did you repeat the experiment (conduct multiple trials) in **Part A** of the investigation?

Review and Reflect

Review

Before this investigation, students may have known that the Earth has a layered structure with crust, mantle, and core. In a sense, students have a mental model of what the Earth's interior is like without knowing much about the *evidence* that underlies this model. This investigation helped them to better understand the nature of the evidence that scientists have used to improve models of the Earth's interior structure. Students should also be beginning to develop a more sophisticated understanding of the nature of these layers, which will be explored in greater depth in **Investigations 3** and **4**. Help your students see that although science investigations provide some answers to scientific questions, they often raise further questions. Spend some time having students talk about these possible questions.

1. In compressional (P) waves, as the wave passes by, the material moves back and forth in the direction of wave travel. In shear (S) waves, the material moves from side to side, perpendicular to the direction of wave travel.

2. The speed of seismic waves increases with depth in the Earth, for somewhat complicated reasons. Because of that change in speed, as the wave moves through the Earth its path is curved by refraction.

3. The speed of seismic waves decreases abruptly at the boundary between the mantle and the core. The only reasonable way to explain that is to assume that the composition of the core is different from the composition of the mantle.

Reflect

Be sure to give students adequate time to review and reflect on what they have done and understood in this investigation. Ensure that all students think about and discuss the questions listed here. Be alert to any misunderstandings and, where necessary, help students to clarify their ideas. The questions here are important in helping students to focus on their understanding of the Earth's interior.

4. Answers will vary depending upon students' initial understandings about the Earth's interior—some students know a lot about the Earth's interior, but others know very little. Students should explain how their understandings have changed, and the evidence or reasons behind the changes. Most likely, students will note that they now know about the different layers, that the core actually has two layers, and that we use earthquake waves to find this out.

5. **Part A** of the investigation was a model of wave speed. **Part C** of the investigation was a model of wave refraction. **Part D** of the investigation used the model of wave refraction from **Part C** to predict how seismic waves might be refracted inside the Earth.

6. Answers will vary. Students should note that they have learned that scientists use their knowledge of how waves behave in different materials and how seismic waves travel, and that they collect evidence from seismographs to make models of the Earth's interior.

Thinking about the Earth System

The overall goal here is that students will have a working understanding of the Earth System by the time they complete eighth grade. It is important that they have time to make any connections they can between what they have done in **Investigation 2** and the Earth System. Discuss with your students their ideas about the relationship between the geosphere and the hydrosphere:

- How was the release of energy from dropping a pebble into water somewhat like an earthquake releasing energy into the ocean?
- What might happen to the water waves that are caused by an earthquake when the water waves reach the shoreline?

Refer students to the diagram on page Pviii of their books as a means of exploring the Earth System further. Ensure that students add their ideas to the *Earth System Connection* sheet.

7. Students should note that the Earth has a layered structure, and that these layers have different thickness and are made of different kinds of material, and that the density of the material increases with depth.

8. Students should note that the energy that is released from the geosphere as an earthquake is transferred to the water in the hydrosphere. This creates waves in the ocean, which spread outward from the part of the ocean above the earthquake.

Thinking about Scientific Inquiry

In **Investigation 2**, your students have been exploring some of the evidence that supports our model of the Earth's interior. As with **Investigation 1**, students' mental models of what the Earth's interior is like may have changed as a result of their experiences, in just the same way that their models of the contents of the mystery bag changed as they gathered new evidence. Students also explored an aspect of the nature of science—the value of conducting repeated tests or trials in an experiment and averaging the results, as well as making decisions about "bad measurements" versus "good measurements." Have them review the **Inquiry Processes** listed on page Px of their text. Ask them to review the explanations given for the inquiry processes and consider these when answering the questions below.

9. Answers will vary. Students will likely discuss what they learned about wave refraction and how this knowledge helped them to change their ideas about what the inside of the Earth is like.

10. We repeated the experiment (conducted 10 trials) to get an average for travel time. It is unlikely that we could get the exact answer on one try. Doing many trials allowed us to average the results and get a value that is more likely to be correct than we would get if we did one trial.

Assessment Tool

Review and Reflect Journal Entry-Evaluation Sheet

Use the general criteria on this evaluation sheet for assessing content and thoroughness of student work. Adapt and modify the sheet to meet your needs. Consider involving students in selecting and modifying the criteria for evaluating their reflections on **Investigation 2**.

Pg 305

Teacher Review

Use this section to reflect on and review the investigation. Keep in mind that your notes here are likely to be especially helpful when you teach this investigation again. Questions listed here are examples only.

Student Achievement

What evidence do you have that all students have met the science content objectives?

_____

Are there any students who need more help in reaching these objectives? If so, how can you provide this?_____

_____

What evidence do you have that all students have demonstrated their understanding of the inquiry processes?_____

_____

Which of these inquiry objectives do your students need to improve upon in future investigations? _____

_____

What evidence do the journal entries contain about what your students learned from this investigation? _____

_____

Planning

How well did this investigation fit into your class time?_____

_____

What changes can you make to improve your planning next time? _____

_____

Guiding and Facilitating Learning

How well did you focus and support inquiry while interacting with students?

_____

What changes can you make to improve classroom management for the next investigation or the next time you teach this investigation? _____

_____

How successful were you in encouraging all students to participate fully in science learning? _____

_____

How did you encourage and model the skills values, and attitudes of scientific inquiry? _____

_____

How did you nurture collaboration among students? _____

_____

Materials and Resources

What challenges did you encounter obtaining or using materials and/or resources needed for the activity? _____

_____

What changes can you make to better obtain and better manage materials and resources next time? _____

_____

Student Evaluation

Describe how you evaluated student progress. What worked well? What needs to be improved? _____

_____

How will you adapt your evaluation methods for next time? _____

_____

Describe how you guided students in self-assessment. _____

_____

Self Evaluation

How would you rate your teaching of this investigation? _____

_____

What advice would you give to a colleague who is planning to teach this investigation? _____

Investigation 2

NOTES

INVESTIGATION 3: FORCES THAT CAUSE EARTH MOVEMENTS

Background Information

Convection Currents

Density differences from place to place in a body of fluid produce convection currents, by which cooler, denser materials move downward and warmer, less dense materials rise upward. These motions happen because of the phenomenon of buoyancy: less dense materials tend to rise to be above more dense material, and vice versa, until the body of fluid is horizontally stratified, with the less dense fluid in the form of a horizontal layer that lies above the more dense fluid. If something acts continuously to produce density differences from place to place, however, the fluid keeps moving in response. Convection then becomes organized into one or more elements, or cells, with some arrangement of rising and sinking regions. Convection is one of the most complicated topics in the mechanics of fluids; the scale, the geometry, and the degree of regularity of the convection patterns varies greatly, depending upon many variables.

Vigorous convection cells come about when a body of fluid is heated from below and cooled from above. The convection your students investigate in this activity is a small-scale and relatively simple example of this. Convection is common in the Earth's atmosphere when the Sun heats the ground surface, which in turn heats the lowermost layer of the atmosphere. Geoscientists have known for decades that the mantle, except for its rigid uppermost part, is in a state of convection. Some regions of the mantle are less dense than other regions at the same depth, mostly because their temperature is higher. The cause of temperature differences is usually attributed to differing concentrations of radioactive materials or to unequal heating from the underlying core, or to the downward movement of originally cool lithospheric plates. The problem of mantle convection is an active area of research in geoscience, and much is still not well understood.

The convective motion of the mantle is associated with the movement of the Earth's lithospheric plates, which your students will learn about in the next investigation. It might seem natural to you that the lithospheric plates move as passive "riders" on top of mantle convection cells. Another effect, however, which is less easy to understand, is probably even more important. Rising hot mantle material beneath the mid-ocean ridges heats the overlying rock, causing it to expand; this is the basic cause of the mid-ocean ridges. There is then a downward slope away from the ridge crest, and the uppermost rigid part of the Earth (called lithospheric plates) has a tendency to slide downhill. This phenomenon is called ridge push. Eventually, as the plates cool, they become denser than the hotter mantle below, and they sink down subduction zones. This phenomenon is called slab pull. These motions of the plates caused by the combination of ridge push and slab pull are thought to be, at least in part, the drivers of the convective motions in the mantle, rather than the other way around!

More Information...on the Web

Go to the *Investigating Earth Systems* web site www.agiweb.org/ies for links to a variety of other web sites that will help you deepen your understanding of content and prepare you to teach this module.

Investigation Overview

In **Investigation 3**, students are asked to consider whether or not the Earth's mantle moves. Students conduct a small-scale, hands-on investigation into the process of convection. Students also observe the teacher demonstrate convection using a heated beaker of water, a cup of oatmeal, and food coloring. Students are asked to consider these two activities as models of how convection operates in the Earth by mapping the elements of their experimental setup onto the layers of the Earth that they have studied in prior investigations. Reading in the **Digging Deeper** section introduces formal terms for the processes and materials involved in convection in the Earth, and introduces some of the "dynamic artifacts" of convection at the Earth's surface—the formation of new crust, volcanoes, and mid-ocean ridges.

Goals and Objectives

As a result of **Investigation 3**, students will have a working understanding of the process of convection, and how convection within the Earth causes plates of the rigid, outer shell of the Earth (lithospheric plates) to move and leads to volcanism and the formation of crust at mid-ocean ridges.

Science Content Objectives

Students will collect evidence that:
1. Convection is a motion in a fluid that results from the fluid being heated from below and cooled from above.
2. Earth has an outer rigid shell (the lithosphere) underlain by a layer of material (the asthenosphere) that is hot and under enough pressure to deform slowly over time.
3. Convection currents, driven by uneven heating within the Earth, are associated with movement of lithospheric plates.
4. New crust is formed where plates spread apart.

Inquiry Process Skills

Students will:
1. Use models to investigate questions about plate movement.
2. Compare aspects of a model with an actual event or phenomenon.
3. Apply knowledge from modeling and model analysis to a new situation.
4. Arrive at conclusions based on data analysis.
5. Communicate findings and results to others.

Connections to Standards and Benchmarks

In **Investigation 3**, students explore convection cells in two different kinds of fluids (syrup and water) and relate these convection cells to the uneven heating and cooling within the Earth's lithosphere and asthenosphere. These observations start them on the road to understanding the National Science Education Standards and AAAS Benchmarks shown on the following page.

NSES Links

• The solid Earth is layered with the lithosphere; hot, convecting mantle; and dense, metallic core.

• Lithospheric plates on the scales of continents and oceans consistently move at the rate of centimeters per year in response to movements in the mantle. Major geological events, such as earthquakes, volcanic eruptions, and mountain building, result from these plate motions.

all shallower

AAAS Link

• The interior of the Earth is hot. Heat flow and movement of material within the Earth cause earthquakes and volcanic eruptions and create mountains and ocean basins. Gas and dust from large volcanoes can change the atmosphere.

Preparation and Materials Needed

Preparation

In **Investigation 3**, your students will conduct a simple experiment, and you will need to conduct a demonstration. The experiment involves placing a layer, about two to three centimeters thick, of a viscous liquid (corn syrup, pancake syrup, or whatever nonflammable fluid you have available) in a small metal can, carefully placing two small pieces of cardboard onto the top of the syrup, placing the can across two bricks, and sliding a lighted candle under the can. In a few minutes, students will see the cardboard moving. This motion stems from the convection of the heated fluid.

If the cardboard pieces get soggy, they tend to sink. To prevent this, you can laminate a manila folder, then cut out 1-cm squares. The top of the candle flame should just reach the bottom of the can. If the bricks are too tall or the candle is too small, the syrup will not heat quickly. If you use cans that once held tuna or cat food, remove all paper and clean the cans thoroughly before use. Prepare for necessary safety precautions. When working with an open flame, students should wear safety goggles, an apron, tie back hair (if necessary), and roll up their shirtsleeves.

For the teacher demonstration, you need to heat a large heatproof beaker of water. As the beaker is heating, and you begin to notice convection in the water, pour in the cup of oatmeal. Try to do this before the water boils. Add the drop of food coloring to the beaker, but ask students to observe before you do this—the convection is demonstrated very quickly, and soon the entire beaker will be colored. Sometimes it is helpful to hold a sheet of white poster board behind the beaker after pouring in the oatmeal and food coloring. Keep it a safe distance from the heat source, but it will provide a background that will make convection easier to see.

Set a good example for your students by following proper laboratory safety precautions. When doing the demonstration, wear safety goggles and an apron, tie back your hair (if necessary), and roll up your sleeves. Make sure that students are sitting a safe distance from the heating beaker. If they must approach the beaker, they should be wearing goggles and aprons.

You may wish to use **Blackline Master** *Our Dynamic Planet* 3.1, Experimental Setup – **Investigation 3** to make an overhead of the experimental setup that you can use when introducing the activity to students. It might be helpful for each student to have a copy of **Blackline Master** *Our Dynamic Planet* 3.2, Convection Cells for **Steps 6 - 8** of the investigation (or you can use the **Blackline Master** to make an overhead transparency to display in the classroom).

Materials

- candle
- small metal can (empty, clean tuna fish can, label removed)
- corn syrup (or pancake syrup)
- two pieces of thin cardboard, 1 cm square   *al foil   or wax paper?*
- two small bricks

For the demonstration (**Step 5** of the investigation), the teacher will need the following materials:

- hot plate
- large clear heatproof beaker (1L)
- cup of oatmeal
- food coloring

INVESTIGATING OUR DYNAMIC PLANET

Investigation 3:

Forces that Cause Earth Movements

Key Question

Before you begin, first think about this key question.

Explore Questions

Does the rock of the Earth's mantle move?

Materials Needed

For this investigation, each group will need:

• candle

• small heat-resistant container

• corn syrup

• two pieces of thin cardboard, I cm square

• two small bricks

• lighter or matches

In the previous investigation you learned that the outer part of the Earth, above the core, is called the mantle. It consists mostly of solid rock. Is it possible that this rock moves?

Share your thinking with others in your class. Keep a record of the discussion in your journal.

Investigate

Design Investigation

1. Set up the experiment as shown on the opposite page. Place a small heat-resistant container where it can receive heat from a candle.

   Pour one centimeter of cold corn syrup into the container.

Key Question

Use the **Key Question** as a brief warm-up activity to elicit students' ideas about whether or not the rock of the Earth's mantle moves. Tell students to write down their ideas in their journal. In addition to choosing either "yes" or "no", students should explain their answer. Why do they think it can move? Why do they think it cannot move? What evidence do they have for either answer? After a few minutes, discuss students' ideas in a brief conversation. Emphasize thinking and sharing of ideas. Avoid seeking closure (i.e., the "right answer"). Make students feel comfortable sharing their ideas by avoiding commentary on the correctness of responses. Record all the ideas that students share on an overhead transparency or on the chalkboard.

Student Conceptions about the Movement of Mantle Rock

Students were told in **Investigation 2** that the mantle is made mostly of solid rock, and that the passage of shear waves through the mantle provides the evidence for this claim. Still, the idea of the mantle being solid takes time for students to understand. Even more challenging is for students to comprehend that solid rock can be deformed. The majority of students will likely argue that solid rock cannot move, or if it moves, cannot move very far.

Answer for the Teacher Only

The solid rock of the mantle can be deformed and flow because it is hot enough and under enough pressure to be deformed. The deformation takes place very slowly, but over millions of years, mantle rock can flow thousands of kilometers. It took time for scientists to come to this understanding. In fact, most scientists rejected the idea of continental drift for years because they did not see how continents could move.

Assessment Tool
Key Question Evaluation Sheet
Remind students to look at the criteria that outline the basic expectations for the warm-up activity. This evaluation sheet emphasizes that you want to see evidence of prior knowledge and that students should communicate their thinking clearly.

About the Photo
The photograph shows a row of volcanoes along a rift zone in Iceland. Active volcanoes provide evidence that the rock of the Earth's mantle moves. Iceland, built almost completely of cooled basaltic lava, is one of the few places in the world where the mid-ocean ridge rises above sea level. Iceland straddles the Mid-Atlantic Ridge, where the North American and Eurasian plates are moving apart at a speed of several centimeters per year.

Investigate

Teaching Suggestions and Sample Answers

Introducing the Investigation

Review **Steps 1 - 4** with students. The investigation seeks to find out if a fluid can flow if heated. The next step (and a big one) is to see if there is any evidence to support their ideas about whether or not the solid rock of the mantle can move. Remind them about safety precautions. Students should not handle the can after it has been heated.

1. If you think it necessary, demonstrate how to set up the experiment.

NOTES

10/5   Used aluminum foil and brown corn syrup. Foil did not get carried away. Even w/ vigorous heating bubbles formed (boiling); bubbles lifted foil.

10/12   used convection activity in NSTA Prj Earth Science Geology (pg 21-). Provides gd illustration of convection systems

Investigation 3

Place two pieces of cardboard so they touch, side by side, on top of the syrup in the center of the container.

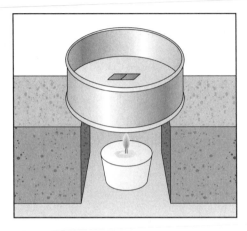
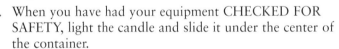

Inquiry

Making Diagrams

Sometimes the best way to show the results of a scientific investigation is by drawing a diagram. Complicated concepts can often be illustrated more easily than they can be explained in words. The diagram should be labeled and described in a sentence or two.

2. Before you light the candle, predict what you think will happen to the cardboard as the corn syrup heats up.

   a) Record your prediction, and also the reason(s) for your prediction.

3. When you have had your equipment CHECKED FOR SAFETY, light the candle and slide it under the center of the container.

   a) While waiting for the corn syrup to heat, draw a side view of the container alongside your prediction and reason(s).

4. Watch carefully as the corn syrup heats up.

   a) Record your observations and interpretations in your journal.

   b) On your "side-view" drawing of the experimental setup, show any movement of the cardboard with solid arrows.

   c) Show the movement of the corn syrup with dashed arrows.

5. Next you will watch a demonstration.

   A clear, heat-proof beaker, two-thirds full with water is placed on a hot plate.

   When the water is simmering (not boiling), one cup of oatmeal will be poured into the beaker. One drop of food coloring will be added. Finally, some sawdust will be added.

Be careful with open flames. Tie back hair and push back or roll up loose clothing. Remove loose jewelry from near your hands. Wear goggles. Allow the syrup to cool completely before touching the heat-resistant container. Clean up spills immediately.

This is a demonstration only. Once the oatmeal is taken into the lab, it must be considered contaminated because it is used in a science activity. Do not consume the oatmeal.

P
23

2. Students should record their prediction and the reason for their prediction in their journals.

   a) Answers will vary. Sample: "I think that the cardboard will move when the syrup is heated because the syrup will boil, and the rising bubbles will move the cardboard."

3. Circulate around the room to make sure that students are following proper safety procedures.

   a) Students are asked to draw and label a diagram. The idea here is for students to think about what is happening to the syrup. Drawing a side view helps them to do this.

4. a) Students should note that the cardboard moves as the syrup heats up. Sometimes the two pieces of cardboard stick together and they move in unison across the surface of the syrup. Other times, the upward movement of syrup spreads the two pieces of cardboard apart.

Teaching Tip

At this point, have students blow out the candle and return to their desks. There is no need for them to handle the can of syrup, which will be hot. While students are writing up their observations, you can begin heating the beaker of water for the demonstration.

---

   b) Students should note any movement of the cardboard by placing arrows on their diagrams.

   c) Students should show the syrup moving from the bottom of the pan to the top of the pan, and then moving horizontally at the top of the syrup.

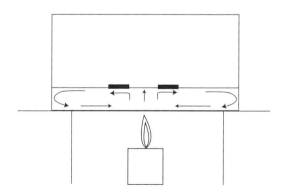

5. Have students read **Step 5**. Tell them that you are going to heat up the beaker of water, and that before the water boils, you are going to add a cup of oatmeal to the water. Ask them to write a prediction about what they think will happen to the oatmeal when it is added, and to give the reason for the prediction as described in **Step 5(a)**.

a) Write a prediction about what you think will happen to oatmeal, food coloring and sawdust when they are added to the water. Give a reason for your prediction.

When the hot plate is turned on and the water begins to warm, carefully observe what happens when oatmeal, food coloring, and sawdust are added.

b) Describe what you observe. Draw a series of diagrams to record your observations.

Evidence for Ideas

6. Below is a simple cross-section diagram of a mid-ocean ridge, showing both what is above and below the Earth's surface.

Think about how the corn syrup model and the oatmeal model behaved, especially the cardboard and sawdust, as the liquid heats up. Compare this to the diagram.

Consider Evidence

a) What evidence can you find from your models that might be similar to what you see in the diagram?

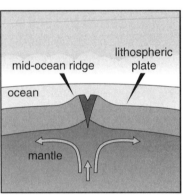

7. What parts or processes within the Earth do you think each of the following parts of your model represent?

Consider Evidence

a) the heat source;

b) the upward movement of water/syrup;

c) the horizontal movement of water/syrup;

d) the separation of the cardboard by syrup;

e) the horizontal movement of the cardboard/sawdust.

8. Hold a whole-class session where each group in turn posts and explains its diagrams to others. Look for similarities and differences and try to reach some overall agreements about the Atlantic's mid-ocean ridge.

Show Evidence

5. a) Sample answer: I think that the oatmeal will sink to the bottom of the beaker, but because heat rises, the oatmeal will rise after it sinks to the bottom.

    b) Students should observe the oatmeal circulate in a convection cell in the beaker. When you add the food coloring, they will observe it circulating or spinning in the water with the oatmeal.

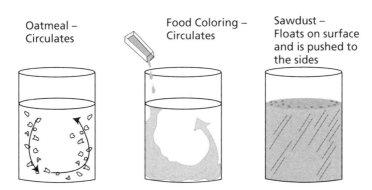

Oatmeal – Circulates

Food Coloring – Circulates

Sawdust – Floats on surface and is pushed to the sides

6. Ask a student to describe the diagram on page P24.

    a) Students should note that the arrows in the mantle show a motion of mantle material like the motion of the oatmeal and syrup. Students should also note that the cardboard moved across the syrup like the lithospheric plate moved across the surface of the syrup. They might also infer that the core is hot, like the candle flame or heat source in the demonstration, and must be heating the mantle from below.

7. Hand students a copy of **Blackline Master** *Our Dynamic Planet* 3.2, Convection Cells, or display an overhead transparency of that diagram.

    a) The interior of the Earth is like the heat source in the model.

    b) The movement of rock in the mantle is like the movement of oatmeal or syrup.

    c) The horizontal movement of the top part of the mantle is like the horizontal movement of the syrup at the top of the heated can.

    d) The separation of the lithospheric plates at the mid-ocean ridge is like the separation of the cardboard pieces in the syrup.

    e) The horizontal movement of the lithospheric plate is like the horizontal movement of the cardboard pieces.

8. It might be a good idea to collect student work, review the work, and make overhead transparencies of several diagrams. When the class next meets, have students present their ideas at the overhead projector.

Assessment Tools

Journal Entry-Evaluation Sheet
Use this sheet as a general guideline for assessing student journals, adapting it to your classroom if desired.

Journal Entry-Checklist
Use this checklist as a guide for quickly checking the quality and completeness of journal entries.

Making Connections...*with History*

The following article provides an overview of the history of plate tectonic theory. Share this with your students' history or social studies teacher for possible connections to scientists of the nineteenth and twentieth centuries. Smith, Michael J., and Southard, John B. (2001) *Exploring the Evolution of Plate Tectonics.* Science Scope Vol.25, No. 1, pp. 46-49.

NOTES

Investigation 3

Investigation 3: Forces that Cause Earth Movements

Digging Deeper

Evidence for Ideas

FORCES THAT DRIVE OUR DYNAMIC PLANET

Convection Cells

Convection is a motion in a fluid that is caused by heating from below and cooling from above. The corn syrup and oatmeal in your investigation were convecting. When a liquid is heated, it expands slightly. That makes its density slightly less. The fluid with lower density then rises up, in the same way that a party balloon filled with helium rises up. With the balloon, you can even feel the upward tug on the string! When the heated liquid reaches cool surroundings, it shrinks again, making its density greater. It then sinks down toward where it was first heated. This circulation, which you observed in the corn syrup, and in the water/oatmeal mixture, is called a convection cell.

key parts
- Hot rising
- cold descending
- Heat at bottom

Convection in the Earth's Mantle

In **Investigation 2** you learned that the Earth's mantle extends down to the hot iron core. It is known that P waves can pass through solids, liquids, and gases, but S waves can go only through solids. Both P waves and S waves go through the mantle, so it must be solid. On the other hand, only P waves go through the core, so its iron must be melted rather than solid. Scientists are now sure that the mantle convects, in the form of gigantic convection cells. How can that be, if the mantle is solid rock?

What defines ridge?
"   " trench
what are plate boundaries

What determines positions of MORs?

As You Read...

Think about:

1. What are the conditions that cause convection cells in a fluid?
2. How can the mantle convect if it is a solid?
3. What is the typical speed of mantle convection?
4. What is the reason for volcanic activity along mid-ocean ridges?
5. What kinds of forces drive sea-floor spreading?

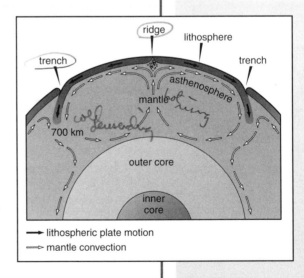

lithospheric plate motion
mantle convection

Digging Deeper

This section provides text and illustrations that give students greater insight into the nature of convection cells, convection in the mantle, the lithosphere and asthenosphere, and the formation of mid-ocean ridges. You may wish to assign the **As You Read** questions as homework to help students focus on the major ideas in the text.

As You Read..

Think about:

1. Convection occurs when a fluid is heated from below and cooled from above. Heating the fluid makes it expand slightly, which makes it less dense, so it rises. When the heated fluid cools at the top of the cell (usually when it approaches or reaches the surface), it contracts or shrinks, becomes more dense, and sinks.

2. The mantle can convect because the mantle rock is just hot enough and under enough pressure to slowly deform. In effect, the mantle rock acts like a liquid on a very long time scale.

3. The typical speed of mantle convection is a few centimeters per year.

4. Hot mantle rock rises in convection cells. As it approaches the surface, some of it melts to form magma, and the magma forms volcanoes on the sea floor.

5. The melting of mantle rock at the top of a rising convection cell provides a supply of magma that forces the lithosphere apart at mid-ocean ridges. The downhill slope of the ridge away from the crest of the ridge adds to the forces that move the new ocean crust away from the ridge.

Assessment Opportunity

You may wish to rephrase selected questions from the **As You Read** section into multiple choice or "true/false" format to use as a quiz. Use this quiz to assess student understanding and as a motivational tool to ensure that students complete the reading assignment and comprehend the main ideas.

Investigation 3

About the Illustration

The illustration on page P25 is a schematic cross section of the Earth. It shows mantle convection cells. The actual patterns of mantle convection are still not very well known, and they are likely not to look exactly like the ones shown in this diagram. Note that mid-ocean ridges form where mantle convection brings material to the surface to create new lithosphere, and that ocean trenches form where mantle convection takes lithosphere back down into the mantle. The lithosphere seems to act like a passenger on a conveyor belt, somewhat like how the cardboard pieces were conveyed along the top of the convecting syrup. Note the value "700 km" in the mantle where the lithosphere is being subducted. That is the about the depth below which the subducted plate loses so much rigidity that earthquakes no longer occur.

NOTES

INVESTIGATING OUR DYNAMIC PLANET

Many materials act like solids on short time scales but like liquids on much longer time scales. If you've ever played with Silly Putty®, you know all about that. Ordinary glass is also a good example. You know that it breaks easily. But if you were to take a long glass rod and hang it horizontally between two supports, it would gradually sag down in the middle. It would have flowed, to take on a new shape, even though it seems like a solid. The Earth's mantle behaves in just the same way. The speeds of flow in the mantle are only a few centimeters per year, but over millions of years of geologic time, that adds up to a lot of movement. Here's a comparison that will give you a good idea of how fast the convection cells in the mantle move: about as fast as your fingernails grow!

The Lithosphere and the Asthenosphere

The outermost part of the Earth, down to a depth of 100 to 200 km in most places, is cooler than the deeper part of the Earth. Because this outermost part of the Earth is relatively cool, it stays rigid, and it does not take part in the convection of the mantle. It is called the lithosphere ("rock sphere"). The lithosphere is made up of the crust and the uppermost part of the mantle. Below the lithosphere is a zone where the mantle rocks are just hot enough and under enough pressure that they will deform and change shape. This zone is right below the lithosphere and is called the asthenosphere. You can think of the lithosphere as a rigid slab that rides on top of the convecting asthenosphere. That is much like the cardboard that rode on top of the syrup in your model. The lithosphere consists of several pieces, each in a different part of the world. These pieces are called lithospheric plates.

Earthquakes exist down to 700 km

Teaching Tip

Early in the era of plate tectonics, it was generally thought that the plates ride passively on the upper limbs of the convection cells. It is now believed that the plates play an active role in maintaining the convection. As an oceanic plate moves away from a spreading ridge, it cools, and as it cools, it becomes denser than the hot mantle below. In a sense, it pulls itself down the subduction zone! This, together with the pull of gravity on the plate because of the downhill slope away from the ridge crest, act to maintain the convective circulation in the mantle. The role of friction exerted on the underside of the plate by the moving mantle material is still a subject of debate, but it is likely not to be as important a factor as the two ("slab pull" and "ridge push") discussed above.

Teaching Tip

You may wish to use the following **Blackline Masters** to make overhead transparencies to use while discussing the **Digging Deeper** reading section:
- *Our Dynamic Planet* 3.2, Convection Cells,
- *Our Dynamic Planet* 3.3, Mantle Convection,
- *Our Dynamic Planet* 3.4, Major Lithospheric Plates and Kinds of Plate Boundaries
        and
- *Our Dynamic Planet* 3.5, Sea-Floor Spreading

Mid-Ocean Ridges

All the Earth's oceans have a continuous mountain range, called a mid-ocean ridge. These ridges are greater than 80,000 km long in total. The Earth's mid-ocean ridges are located above rising currents in mantle convection cells. You might think that the ridges are formed by the upward push of the rising mantle material, but that's not the reason. The ridges stand high because they are heated by the hot rising material. Like most materials, rocks expand when they are heated.

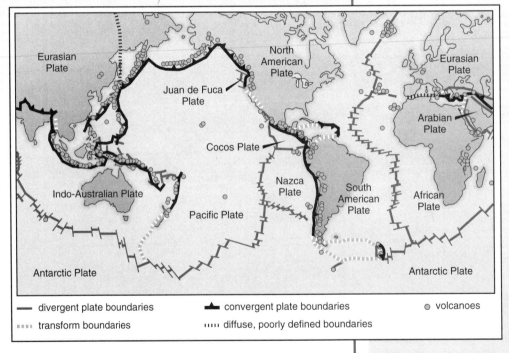

- — divergent plate boundaries
- •••• transform boundaries
- ▲— convergent plate boundaries
- ıııı diffuse, poorly defined boundaries
- ⊙ volcanoes

As the hot mantle rock rises up toward the mid-ocean ridge, some of it melts, to form molten rock called magma. The magma is less dense than the surrounding rocks, so it rises up to form volcanoes along the ridge. The reason for the melting is not obvious. As the rock

Selective melting

Are these volcanoes cones?

About the Illustration

The map on page P27 shows the locations of the world's major lithospheric plates. Note the large number of volcanoes that are located along convergent boundaries (places where one plate is subducted beneath another). Note that two other kinds of plate boundaries are shown on the map—convergent boundaries and transform boundaries. Students will learn about these two kinds of boundaries in **Investigation 4**. Volcanoes are numerous along divergent boundaries (mid-ocean ridges) as well, but most are in the deep ocean, and can be observed only by means of deep-diving submersibles.

Investigation 3

INVESTIGATING OUR DYNAMIC PLANET

[handwritten margin notes top right] PT $\frac{L}{S}$ P →

[handwritten left margin] Partial melting (ex cookie)
Tm ∝ P, T, water
chem

rises, it stays at about the same temperature, but the pressure on it decreases, because there is less weight of rock above it. It's known, from laboratory experiments, that the melting temperature of most rocks decreases as the pressure decreases. That's why some of the rising rock forms magma.

When the magma reaches the surface of the ridge, it solidifies to form a rock called basalt. That's how new crust is formed in the oceans. As soon as the new crust is formed, it moves away from the crest of the ridge. The movement is partly from the force of the moving mantle *[handwritten]* ↳ below. It is also partly because of the downhill slope of *[handwritten]* 2 gravity the ridge away from the crest. The movement of new *[handwritten]* (>P) oceanic crust in both directions away from the crest of a mid-ocean ridge is called sea-floor spreading. *[handwritten]* Divergent plate boundary.

[handwritten left margin] Ridge push
Slab pull
Help drive convection
(pg 113)

[handwritten left margin] revised.
See ies
web sites

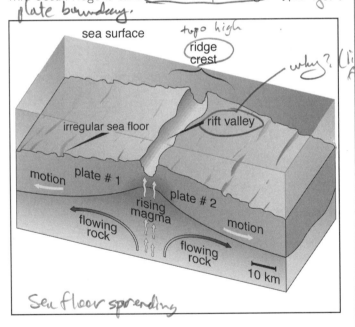

sea surface
ridge crest
irregular sea floor
rift valley
motion
plate # 1
rising magma
plate # 2
flowing rock
motion
flowing rock
10 km

[handwritten diagram notes] topo high   why? (like French doors)

[handwritten] Sea floor spreading

[handwritten bottom] So what drives convection?
① Heat from core (radioactive decay & latent heat of impact)
② Movement of plates: ridge push   slab pull

About the Illustration

The illustration on page P28 shows sea-floor spreading at a mid-ocean ridge. Note how the ridge slopes away from the crest of the ridge. New lithosphere moves down the slope of the ridge under the force of gravity, and is also spread apart by convection in the mantle below.

Ophiolite Suites

pillow lavas
dikes
gabbro
magma

sea

gabbro

meta &
alt

peridotites

Model

- Ocean floor is basalt & cooled surface layer of convection in upper mantle

- Explains origin & design of ocean floors.

- Clues: Iceland, ophiolite (cyprus, Newfoundland Caribbean)

- see pg. 14 of hour ted

- Driven by non-solar source of heat.

Evidence
for Ideas

Review and Reflect

Review

1. Think back on the **Key Question**: "Does the rock of the Earth's mantle move?" Answer this question again, based on what you learned in this investigation.

2. Why is the mid-ocean ridge made of volcanic rock?

3. Is the rock that makes up the mid-ocean ridge young or old? Explain your answer.

Reflect

4. Give at least one example of where a convection cell could be formed. Use an example different from any that are presented in this investigation. Describe why the convection occurs.

5. Do you think that a convection cell would be formed in a fluid that is heated at the surface rather than at the bottom? Explain your answer.

Thinking about the Earth System

6. What connections did you discover between convection in the mantle (geosphere) and the oceans (hydrosphere)?

7. How does convection in the Earth's mantle help to shape the geosphere?

8. Describe the flow of energy in the geosphere. Think about the movement of heat energy during mantle convection.

Thinking about Scientific Inquiry

9. a) What hypothesis did you form in this investigation?

   b) Was your hypothesis proven right or wrong by your investigation?

   c) If your hypothesis was wrong, are the results of the investigation still valid? Explain.

10. Explain why a diagram was a good way to show your observations in this investigation.

Review and Reflect

Review

Give your students ample time to review what they have learned from **Investigation 3**. This has been a laboratory investigation, and your students may need help in relating their findings to the real world.

The answers provided below are for you, the teacher. It is not expected that your students will answer with the same level of sophistication. Use your knowledge of the students as well as the standards set by your school district to decide what answers you will accept. In student answers, look for evidence of an understanding of the processes involved, as well as for any misconceptions that still remain. Encourage students to express their ideas clearly and to use correct science terminology where appropriate.

1. Students will most likely answer "yes," and should provide additional insight that it moves very slowly, about as fast as a fingernail grows. In greater detail, the rock of the mantle moves slowly, as a consequence of the existence of large-scale convection in the mantle. The speed of a given parcel of mantle rock ranges from a few centimeters per year to as much as 20 cm per year. Over millions of years, mantle rock can move great distances. Geoscientists say that the mantle rock flows as a plastic. In materials science, a plastic is a solid material that does not flow until the deforming force builds up to a certain value, called the yield strength of the material, and then, at greater deforming forces, it flows as a fluid. The viscosity ("stiffness") of mantle rock is extremely high, however, far higher than even the most viscous fluids with which we are familiar in everyday life.

2. Students should answer that the mid-ocean ridge is made of volcanic rock because the ridge forms above where there are rising currents in mantle convection cells. In greater detail, as the mantle rock rises, the pressure exerted on it by the column of overlying rock decreases. The rock cools only very slightly, however, because transfer of heat by conduction is a very inefficient process. It's known, from laboratory experiments, that the melting temperature of mantle rock decreases with decreasing pressure. Therefore, as the rock rises, and the pressure on it decreases without the temperature decreasing correspondingly, the local melting temperature is reached, and some of the rock melts to form a magma. The magma then rises buoyantly toward the ridge crest, because it is less dense than the surrounding mantle rock. It comes out on the sea floor as submarine lava, and it is cooled rapidly to form volcanic rock.

3. Rock that makes up the mid-ocean ridge is very young. It is new ocean crust.

Reflect

Give your students ample time to reflect on the nature of the evidence they have generated from their investigations. Again, help them see that evidence is crucial in scientific inquiry.

4. Answers will vary. A convection cell can form in a room that is heated. The heater is usually along the baseboard of the wall. Air warmed by the heater expands, becomes less dense, and rises toward the ceiling. At the ceiling it cools, contracts, becomes more dense, and sinks. The resulting circulation is one kind of convection cell.

   A puffy cumulus cloud or a large thundercloud is another good example of convection. Air near the ground is heated by the Sun, becomes less dense, and rises. Sometimes it rises far enough for water vapor to condense to form a cloud. The air keeps rising to some altitude, perhaps high enough for rain or snow to form in the cloud. The descending part of the convection cell is represented by the slowly and broadly sinking clear air in the vicinity of the cloud.

5. A convection cell would not form in a liquid that is heated from above. The heated liquid would expand slightly and become less dense than the liquid below. Less dense liquid could not sink below more dense liquid, so the warmed liquid would remain at the top of the container. One would not expect convection to occur in a lake on a warm summer day, for example.

Thinking about the Earth System

Now that your students have explored the forces that cause Earth movements, and the nature of large-scale convection in the Earth's mantle and the associated movements of lithospheric plates, help them to make connections with other Earth systems. There are many such connections. Plate movements give rise to mountains, which change the patterns of weather and climate over large areas, as well as influencing the fauna and flora of those areas. Plate movements change the configuration of the world's major oceans, with profound consequences for weather and climate, as well as for marine organisms. These are just a few of the connections between plate movements and other aspects of the Earth system. Ensure that students add their ideas to the *Earth System Connection* sheet.

6. The topography of the ocean floor depends, in large part, upon the sea-floor spreading at the mid-ocean ridges and subduction at convergent margins. The deepest parts of the ocean are in the trenches that are formed where one plate bends downward to be subducted under another plate. The configuration of the oceans basins is a consequence of continental drift brought about by plate tectonics.

7. Mantle convection leads to volcanoes and the formation of mid-ocean ridges, which are long, linear mountain ranges on the sea floor. Sea-floor spreading pushes the new ocean crust apart, which changes the shape of ocean basins. Mountain ranges are formed along subduction zones and at continent–continent collision zones.

8. Answers will vary. Students should note that heat is transferred from the top of the Earth's core to the base of the mantle. The heat causes mantle convection to occur, and as material rises in convection cells the heat is transferred to the Earth's surface. As heat escapes from the oceanic crust, the crust cools and sinks back down into the mantle. The entire process acts to transfer heat from the core to the Earth's surface (and eventually to the atmosphere and space). *& "cold" to mantle*

Thinking about Scientific Inquiry

In this investigation, students have been using maps as tools for investigation. They have also been exploring questions for inquiry and collecting and reviewing information. Help them to understand how they have used these processes in a scientific way by reviewing the list of inquiry processes shown on page Px in the Student Book.

9. a) Students formed a hypothesis about what would happen to oatmeal when it was placed in a beaker of water while the water was being heated.

   b) Answers will vary. Check for consistency between their hypothesis and their description of the outcome of the experiment.

   c) A hypothesis should not affect the outcome of an experiment. If the experiment does not confirm the hypothesis, the hypothesis should be modified, not the experiment.

10. A diagram helped to demonstrate the movement of material in a convection cell, or the changes that took place over time. It is sometimes much easier to use a diagram to show what happens than to explain it in words. Labeled diagrams are even better than unlabeled diagrams.

Assessment Tool
Review and Reflect Journal Entry-Evaluation Sheet
Use the general criteria on this evaluation sheet for assessing content and thoroughness of student work. Adapt and modify the sheet to meet your needs. Consider involving students in selecting and modifying the criteria for evaluating their reflections on **Investigation 3**.

Teacher Review

Use this section to reflect on and review the investigation. Keep in mind that your notes here are likely to be especially helpful when you teach this investigation again. Questions listed here are examples only.

Student Achievement

What evidence do you have that all students have met the science content objectives?

_____

Are there any students who need more help in reaching these objectives? If so, how can you provide this?_____

_____

What evidence do you have that all students have demonstrated their understanding of the inquiry processes?_____

_____

Which of these inquiry objectives do your students need to improve upon in future investigations? _____

_____

What evidence do the journal entries contain about what your students learned from this investigation? _____

_____

Planning

How well did this investigation fit into your class time?_____

_____

What changes can you make to improve your planning next time? _____

_____

Guiding and Facilitating Learning

How well did you focus and support inquiry while interacting with students?

_____

What changes can you make to improve classroom management for the next investigation or the next time you teach this investigation? _____

_____

How successful were you in encouraging all students to participate fully in science learning?_____

_____

How did you encourage and model the skills values, and attitudes of scientific inquiry? _____

_____

How did you nurture collaboration among students?_____

_____

Materials and Resources

What challenges did you encounter obtaining or using materials and/or resources needed for the activity? _____

_____

What changes can you make to better obtain and better manage materials and resources next time? _____

_____

Student Evaluation

Describe how you evaluated student progress. What worked well? What needs to be improved? _____

_____

How will you adapt your evaluation methods for next time?_____

_____

Describe how you guided students in self-assessment. _____

_____

Self Evaluation

How would you rate your teaching of this investigation? _____

_____

What advice would you give to a colleague who is planning to teach this investigation? _____

_____

Investigation 3

NOTES

INVESTIGATION 4: THE MOVEMENT OF THE EARTH'S LITHOSPHERIC PLATES

Background Information

Plate Boundaries

At the heart of the concept of plate tectonics is the idea that large segments of Earth's lithosphere (consisting of the crust and the rigid uppermost part of the mantle) move relative to one another. New lithosphere is continuously created at the mid-ocean spreading ridges. This lithosphere consists of a thin layer of ocean crust above and a layer of rigid mantle below. This rigid lithosphere sits atop a partially molten, less rigid layer called the asthenosphere ("weak sphere"). The asthenosphere is hot enough to deform and flow, while the lithosphere is not. The boundary between the lithosphere and asthenosphere is thus really a temperature boundary. As the lithosphere moves away from its place of origin at the spreading ridge the mantle underneath it cools, and the temperature boundary that separates the lithosphere and asthenosphere deepens. As this happens the lithosphere becomes thicker by progressive downward rigidification of underlying mantle asthenosphere.

Because the Earth is not becoming larger, the addition of surface material in one place must be accompanied by an equal loss of surface material somewhere else on the planet. This loss takes place at subduction zones. Also, in certain places two lithospheric plates slide parallel to each other. There are thus three basic kinds of plate boundaries: divergent boundaries at spreading ridges; convergent boundaries at subduction zones and sites of continent–continent collision; and transform boundaries where plates are sliding past each other.

Divergent boundaries exist where two plates are moving apart. This occurs where convection currents in the mantle rise toward the surface and spread apart. Accordingly, divergent plate boundaries are characterized by igneous rocks that originate from the partial melting of mantle rock. These rocks include an array of dark-colored igneous rocks that are either erupted at the surface (basalt) or solidify within the Earth (gabbro). Igneous activity and shallow-focus earthquakes are common occurrences at mid-ocean ridges. As the plates move away from each other, a geographic feature known as a rift valley is produced. Rift valleys are usually found atop mid-ocean ridges such as the Mid-Atlantic Ridge. Two examples of divergent boundaries on land are the East African Rift Zone and Thingvellir in Iceland.

Convergent boundaries exist where two plates move toward each other. The processes associated with convergent boundaries differ depending upon the nature of the lithosphere involved. Crustal material can be continental (granitic) or oceanic (basaltic). Granitic rock is less dense than basaltic rock and therefore cannot subside into the higher-density mantle. As a result, convergent boundaries can be divided into three subtypes, each of which has characteristic features.

If both of the plates are capped by oceanic crust, one of the plates moves downward under the other plate at some angle, which ranges from rather shallow in the case of fast convergence to almost vertical in the case of slow convergence. This pattern of convergent

Investigation 4

motion is called subduction. A deep trench is formed on the sea floor at the point where the one plate bends downward under the other. Earthquakes occur along the subducting plate, ranging from shallow-focus earthquakes near the trench to deep-focus earthquakes at depths as great as several hundred kilometers. The earthquakes are the result of stresses within the plates themselves, in response to pushing and pulling of the plate, rather than to slippage between the subducting plate and the enclosing mantle. Also, an arcuate (in the shape of an arc) chain of volcanic islands, called a volcanic island arc, develops above the subduction zone by melting of some of the mantle rock overlying the subducting plate, as the subducting plate releases some of its water content. The Aleutian trench and island arc is an excellent example of a modern subduction zone of this kind, which could be called an ocean–ocean subduction zone.

If one plate is capped by oceanic crust and the other is capped by much thicker and less dense continental crust, the subducting plate is always the one carrying oceanic crust, because it has a higher density. A trench develops just offshore of the continental plate. The pattern of earthquakes is similar to that associated with subduction of an oceanic plate beneath another oceanic plate, but the earthquakes occur beneath the continent and the volcanism forms a chain of volcanic mountains inboard of the margin of the continent. The subduction zone along the west coast of South America is an excellent example of a modern subduction zone of this kind, which could be called an ocean–continent subduction zone.

The final possibility is the convergence of two plates, both of which carry continental crust.

Because continental crust is much less dense than the mantle, neither of these plates is subducted. Instead, one is pushed horizontally beneath the other for some distance, until the friction forces along the boundary between the two plates builds up to be greater than the compressive force that is pushing the plates together. The thickening of the lithosphere in this way results in very high land elevations over very large areas. Compressive fault movements of large masses or sheets of continental crust result in the formation of high mountains at the site of the continent–continent collision. Although volcanic activity is not common, earthquakes are common and can occur far from the plate boundary beneath either plate. The outstanding modern example of continent–continent collision is the collision of the Indian Peninsula with the main mass of southern Asia, which has resulted in the Himalayas mountain chain and the Tibetan Plateau. Old mountain belts, like the Appalachians, are interpreted by geoscientists to have formed from a continent–continent collision earlier in geologic time.

Transform faults exist where two plates slide past each other in parallel but opposite motion. A look at any map of mid-ocean ridges shows that the ridges are offset by numerous transform faults. Movement along transform faults generates frequent earthquakes but usually no volcanic activity. The San Andreas Fault in California is an unusually long transform fault.

More Information...on the Web

Go to the *Investigating Earth Systems* web site www.agiweb.org/ies for links to a variety of other web sites that will help you deepen your understanding of content and prepare you to teach this module.

Investigation Overview

Students begin by hypothesizing about how mountains, volcanoes, and earthquakes might occur where lithospheric plates meet. They then model the collision of two plates (plate convergence) using simple materials. Students relate their observations of the model to a world map of lithospheric plates and then create their own models of other kinds of plate boundaries (divergent boundaries and transform boundaries). Text in the **Digging Deeper** reading section emphasizes the forces that lead to mountain building, volcanism, and earthquakes at various kinds of plate boundaries.

Goals and Objectives

As a result of **Investigation 4**, students will develop a better understanding of the nature of plate tectonics and the features and events that result from plate interactions, and will improve their ability to use models to conduct scientific inquiry.

Science Content Objectives

Students will collect evidence that:
1. The Earth's crust consists of thick, less dense continental crust and thin, more dense oceanic crust.
2. The lithosphere is not one continuous piece, but instead exists as large and small pieces or plates.
3. Plates can be moving apart from one another, moving toward one another, or sliding past one another.
4. Plates with ocean crust are more dense and slide under plates with continental crust when they converge.
5. Earthquakes, mountains, and/or volcanoes can often occur at the boundaries between plates.

Inquiry Process Skills

Students will:
1. Use models to investigate a science question.
2. Make predictions about the outcome of the modeling events.
3. Collect data from the models.
4. Compare data from the models.
5. Use knowledge from a model data analysis to interpret maps.
6. Share findings with others.

Connections to Standards and Benchmarks

In **Investigation 4**, students model what happens at plate boundaries. These observations will start them on the road to understanding the National Science Education Standards and AAAS Benchmarks shown on the following page.

NSES Links

- The solid Earth is layered with the lithosphere; hot, convecting mantle; and dense, metallic core.

- Lithospheric plates on the scales of continents and oceans consistently move at the rate of centimeters per year in response to movements in the mantle. Major geological events, such as earthquakes, volcanic eruptions, and mountain building, result from these plate motions.

- Landforms are the result of a combination of constructive and destructive forces. Constructive forces include crustal deformation, volcanic eruptions, and deposition of sediment, while destructive forces include weathering and erosion.

- Models are tentative schemes or structures that correspond to real objects, events, or classes of events, and that have explanatory power. Models help scientists and engineers understand how things work. Models take many forms, including physical objects, plans, mental constructs, mathematical equations, and computer simulations.

AAAS Links

- The interior of the Earth is hot. Heat flow and movement of material within the Earth cause earthquakes and volcanic eruptions and create mountains and ocean basins. Gas and dust from large volcanoes can change the atmosphere.

- Some changes in the Earth's surface are abrupt (such as earthquakes and volcanic eruptions) while other changes happen very slowly (such as uplift and wearing down of mountains). The Earth's surface is shaped in part by the motion of water and wind over very long times, which act to level mountain ranges.

- Models are often used to think about processes that happen too slowly, too quickly, or on too small a scale to be observed directly, or that are too vast to be changed deliberately, or that are potentially dangerous.

Preparation and Materials Needed

Preparation

You can save time during class by preparing the cardboard models for **Part A** of the investigation ahead of time. At the very least, pre-cut the cardboard materials (cereal box cardboard and corrugated cardboard are needed) to the dimensions specified in the diagrams page P31 of the Student Book. **Part B** of the investigation is open-ended, and the materials needed for this part of the investigation will depend upon what ideas your students generate. Encourage students to develop models that require readily available materials or materials that can be brought from home.

Materials

Part A: Modeling Plate Convergence
- scissors
- thick corrugated cardboard
- thin cardboard (like a cereal box)
- duct tape
- foaming shaving cream
- metric ruler

Part B: Modeling Plate Boundaries
Materials to be selected by teacher and students

INVESTIGATING OUR DYNAMIC PLANET

Investigation 4:

The Movement of the Earth's Lithospheric Plates

Key Question

Before you begin, first think about this key question.

What happens where lithospheric plates meet?

To think about what might happen where plates meet, do these simple demonstrations. Place your hands together palms down, in front of you with your fingers pointing forward and the sides of your thumbs touching. What do you think happens when plates move toward one another? (Push your hands together.) What do you think happens when plates move away from each other? (Move your hands apart.) What might happen when plates slide past one another? (Slide your hands past one another.)

Share your thinking with others in your class. Keep a record of the discussion in your journal.

Materials Needed

For this investigation your group will need:

- scissors
- thick corrugated cardboard
- thin cardboard like a cereal box
- duct tape
- foaming shaving cream
- metric ruler

Investigate

Part A: Modeling Plate Convergence

1. In your group, discuss what you expect would happen when two plates move toward each other. Below are some questions to help your discussion.

Key Question

Use the **Key Question** as a brief warm-up activity that draws out students' ideas about the topic explored in **Investigation 4**. This question is designed to find out what your students know about what happens when lithospheric plates meet to set the stage for inquiry.

Write the **Key Question** on the chalkboard or overhead transparency. Encourage students to try the simple model of plate interactions, such as the ones described in the text. Show students how to do this.

Assessment Tool

Key Question Evaluation Sheet
Use this evaluation sheet to help students understand and internalize basic expectations for the warm-up activity.

Student Conceptions of Events at Plate Boundaries

Investigation 3 helped students to think about what happens at divergent plate boundaries (volcanism). **Investigation 4** encourages students to deepen their understanding of plate tectonics by exploring two additional kinds of plate boundaries and the difference between oceanic lithosphere and continental lithosphere. From the simple demonstration, students are likely to suggest that mountains form where plates come together but will probably not be aware of subduction and the production of magma and volcanic activity at convergent plate boundaries. Students may also not realize that what happens at convergent plate boundaries depends upon the kind of lithosphere (oceanic or continental) that the plates are carrying. Some students will be aware of the San Andreas Fault and know that it is an example of what happens when plates slide past one another, but they may have limited understanding of the nature of faulting and the production of earthquakes. Most middle school students associate earthquakes with volcanoes (and volcanoes and mountains). This investigation will help them to better understand why earthquakes and volcanoes do not always occur together.

(handwritten margin notes)
what type?
1 ocean plate/
2 ocean plate
3 ocean/continent
4 continent/cont
5 ocean/spread center
6 cont/spread center
7/ Transform/ transform
8) continental breaking

Answer for the Teacher Only

Note: the intent of this question was to ask what happens where two adjacent lithospheric plates converge, or move toward one another, not when they "meet" in the sense of simply being in contact with one another. When two plates that carry ocean crust converge, one is subducted beneath the other. When a plate carrying oceanic crust converges on a plate carrying continental crust, the former is always subducted beneath the latter. When two plates carrying continental crust converge, one slides horizontally under the other, but it is not subducted, because continental crust is much less dense than the mantle. Eventually, the continent–continent collision ceases because the friction forces build up to prevent the one plate from under-running the other plate any farther.

About the Photo

The Andes Mountains resulted from the convergence of the Nazca Plate and the South American Plate. The oceanic lithosphere of the Nazca Plate is denser than the continental lithosphere of the South American Plate. As a result, the oceanic lithosphere descends (is subducted) into the mantle. Earthquakes and volcanoes are very common along subduction zones.

Investigate

Teaching Suggestions and Sample Answers

Introducing Part A of the Investigation

Tell students that they are going to use a simple model to explore what might happen when two lithospheric plates meet. Before they start the activity, they need to spend some time in their groups thinking about what the model represents. Remind students that in **Investigation 3** they focused on what happens when mantle convection cells bring mantle rock toward the surface (mid-ocean ridges form and plates spread apart). The first part of **Investigation 4** focuses on what happens when plates meet.

1. Encourage students to discuss and answer the three questions in **Step 1** in their groups, and to record their responses in their journals. Circulate among groups to respond to students' queries. Remind students that the questions ask for their ideas, and that it is okay to speculate at this stage in the investigation.

Plate - Plate Interactions (geometries)

		Types of Mts
1) MOR (divergent)	Ex MAR	axial ridge
2) Conv ocean/ocean	Marianas Indonesia Japan	Volcanic arc
3) conv ocean/cont	Juan de Fuca / Cascades Pacific / Andes	Volc chain (Cascade + Andes)
4) Continental breakup (divergent)	Africa / SA	?
5) Convergent cont/cont	India / Asia	Himalayas
6) Transform	San Andreas	Yes (at irregularities in fault)
7a) Convergent oceanic/MOR		Volc arc
7b) " cont/MOR		Volc chain

NOTES

Investigation 4: The Movement of the Earth's Lithospheric Plates

Explore Questions

Using what you already know about mountains, volcanoes, and earthquakes, work together to find the most reasonable answers to these questions. Record your answers in your journal.

see Edible Tectonics UCAR Windows site

a) How might mountains form there?

b) Might volcanoes develop where plates meet? If so, why?

c) Could there be earthquakes where plates collide? Why?

Design Investigations

2. You are now going to model what happens when two plates move toward one another.

With scissors, cut out two pieces of corrugated cardboard. Make each piece about 8 cm wide and about 20 cm long.

Also cut out one piece of cereal-box cardboard, about 8 cm wide and about 15 cm long.

Assemble the pieces of cardboard as shown in the diagram.

When you are ready to run your model, you will be squirting some shaving cream onto the cereal-box cardboard to make a layer about 3 mm thick.

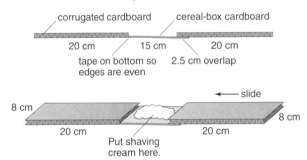

Conduct Investigations

3. You have set up a model of what happens when two plates move toward each other. The process you are modeling is called plate convergence. (The motion of two things toward one another is called convergence.)

The two pieces of corrugated cardboard represent continental plates, and the cereal-box cardboard represents an oceanic plate. The shaving cream represents ocean-floor sediment.

a) Predict what you think will happen when you push the two pieces together until one of the pieces of corrugated cardboard has moved 5 cm beneath the other piece of corrugated cardboard. Also record the reason(s) for your prediction.

a) Answers will vary. Students will likely respond that when two plates move toward one another, mountains get pushed up because the plates get squeezed together.

b) Answers will vary. Students may note that when the two plates meet, one plate might go deeper into the mantle, melt, and allow magma to form.

c) Answers will vary. Students are likely to suggest that yes, earthquakes are likely to happen where plates collides because the force of the collision will break rock layers.

2. As suggested in the preparation section of this guide, you can save time in class by pre-assembling (or pre-cutting) the experimental setup described in this step.

3. Check with students to ensure that they understand what they are modeling. Point out the location of the Nazca Plate (oceanic lithosphere) and the western edge of the South American Plate (continental lithosphere) shown on page P33 of the Student Book as an example of the plates that they are modeling. The models will enable students to see a map view as well as a cross-section view of the "colliding plates."

a) Most likely students will note that they expect the shaving cream to "pile up" as the two "plates" are moved toward one another. Students may also predict that some of the shaving cream will slide underneath the corrugated cardboard.

Teaching Tip

Investigation 4 has students using models to conduct scientific investigations. It might be useful again to discuss the scientific idea of modeling real-world processes that are too difficult to observe directly, or on a scale too large to be changed deliberately, or too dangerous. Students should begin to appreciate how useful models are as a means of observing and testing in scientific inquiry. Help students to see that models must represent as realistically as possible (or as feasibly as possible) what is being replicated.

Use **Blackline Master** *Our Dynamic Planet* 4.1, Experimental Setup – **Investigation 4, Part A** to make an overhead of the experimental setup.

Assessment Tools

Journal Entry-Evaluation Sheet
Use this sheet as a general guideline for assessing student journals, adapting it to your classroom if desired.

Journal Entry-Checklist
Use this checklist as a guide for quickly checking the quality and completeness of journal entries.

b) Draw a side-view diagram of your prediction. On this drawing, use arrows to show the direction of plate movements.

4. Make a data table to record your observations.

Use this example or design your own.

Collect & Review

Use Mathema

MEASUREMENTS OF CONVERGENT PLATE MOVEMENT			
Distance plates moved together	Shape (drawing)	Height (cm)	Width (cm)
2.5 cm			
5.0 cm			
7.5 cm			
10.0 cm			
12.5 cm			
15 cm			
17.5 cm			

5. When you are ready, apply the layer of shaving cream, as described in **Step 2** above.

Now push the two pieces of cardboard together slowly, 2.5 cm at a time. Let the model run until one of the corrugated-cardboard continents has moved under the other corrugated-cardboard continent about 5 cm.

a) Describe what happens as the pieces of cardboard move toward one another.

Evidence for Ideas

b) In your notebook, make a sketch of what the model looks like after the two continents have collided and the one has moved underneath the other.

c) How did your results compare to your predictions?

6. Now look at this map, which shows the plate boundaries. The arrows show the direction the plates are moving.

b) A sketch representing a sample prediction is provided below.

SIDE VIEW

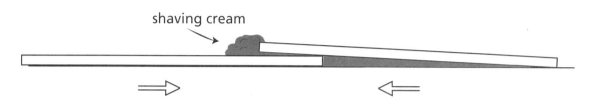

shaving cream

4. The data table provided in the Student Book is just an example. Students often copy such tables into their journals exactly to scale without thinking about the fact that they need space to record their observations. Make sure that students realize that they need to have enough room to record what they observe (in this case, to make sketches in the cells of the data table labeled "Shape (drawing)". One way to do this is to write down the headers of columns and fill the cells as they go, rather than draw the entire data table.

5. If you are concerned about students getting "carried away" with the shaving cream, you can spray the shaving cream onto the models for them. You want a thin, even film of shaving cream, not a thick pile.

   a) Students should note that the layer of shaving cream gets thicker as the cardboard slides, and that some shaving cream slides below the corrugated cardboard.

   b) Check to make sure that sketches are labeled properly and are legible.

   c) Students should make a comparison. If students did not observe what they predicted, they should explain their results.

6. Use **Blackline Master** *Our Dynamic Planet* 4.2, Relative Motions of Major *see* Lithospheric Plates to make an overhead transparency of the figure on page P33 of the Student Book.

Investigation 4

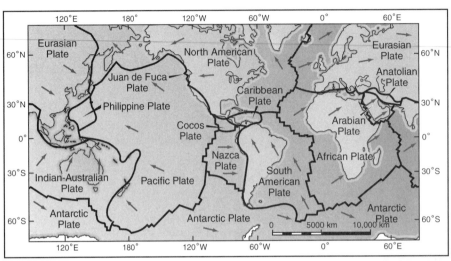

Look closely at those points where an oceanic plate is touching a continental plate. A good example is South America, where the Nazca Plate and the South American Plate are moving toward each other.

a) Where else can you find oceanic plates and continental plates moving toward each other?

b) At a convergent plate boundary, two plates are moving toward each other. How do you think your model might relate to convergent plates like these?

c) What kind of pattern would you expect to see in the locations of earthquakes under the continental plate of South America?

d) Note the places where you think earthquakes or volcanoes might occur.

7. Share your ideas with your group and the rest of the class.

6. a) Examples of where oceanic plates and continental plates are moving toward one another include the Juan de Fuca Plate (oceanic) and the North American Plate in the northwest United States (continental), and the Nazca Plate (oceanic) and the South American Plate (continental) along the west coast of South America. There are many other examples around the world, especially around the western rim of the Pacific Ocean.

    b) The corrugated cardboard of the model represents a continental plate, and the thin cereal box cardboard represents an oceanic plate. As the oceanic plate converges with the continental plate, the oceanic plate slides under the continental plate. Eventually, the two continental plates meet.

    c) Earthquakes are caused by the downward movement of one plate under another. Students might hypothesize, correctly, that (1) there would be a belt or zone of earthquakes parallel to the subduction zone, with one edge located about where the oceanic plate dives down at the edge of the continental plate, and with the other edge located somewhere inland on the continent, and (2) that the depth of the earthquakes increases from very shallow near the former edge of the belt and very deep near the latter edge of the belt.

    d) This question calls for some reasoned speculation. Because the shaving cream is soft and also tends to "lubricate" the contact between the sliding cardboard, students may not fully understand that rocks deform and break as they are compressed together or as one plate is forced beneath another. Accept all reasonable responses. Earthquakes would form at the boundary between the oceanic and continental plates, and volcanoes would form on the continental plate.

        Students should be aware that where an oceanic plate is in contact with a continental plate, there is likely to be subduction of the oceanic plate beneath the continental plate, and that both earthquakes and volcanoes are associated with such plate boundaries. Some details on the earthquakes are given in the answer to **Step 6 (c)** above. The volcanoes are thought to be the result of partial melting of mantle rock just above the downgoing plate at a certain depth (of the order of 100 km), so, because of the downward angle of subduction, the volcanoes are located on the continent well inland of the line of subduction at the margin of the continent. Students might also be aware that the two plates might be sliding parallel to one another, as along the San Andreas Fault in California. That causes earthquakes but no volcanoes.

7. Encourage students to share their ideas in a class discussion and to cite the evidence or reasons for their ideas.

Investigation 4

Teaching Tip

You may suggest that students do additional research to supplement their findings. From the standpoint of scientific inquiry, this is very important. Students need to understand that scientists do not simply engage in experiments. They also spend a great deal of time searching for relevant information to inform their investigations. The *Investigating Earth Systems* web site at www.agiweb.org/ies/ provides links to relevant web sites.

NOTES

Investigation 4

Part B: Modeling Plate Boundaries

1. In **Part A** of this investigation you modeled a convergent plate boundary between continental crust and oceanic crust.

   Read the **Digging Deeper** section that describes other plate boundaries.

   Choose one type of plate boundary to model. You may wish to model a convergent plate boundary between continental and oceanic crust as in **Part A**, using different materials.

   a) Record in your journal the boundary you have decided to model. Write your investigation in the form of a question.

   b) Based on the information you have, predict what you think will happen at the boundary and explain why.

2. In your group discuss the best way to model the boundary you have selected. Consider the materials that are available to you.

   a) List the materials you plan to use. Explain why you chose the materials that you did.

   b) Outline the steps that are required to set up and run your model.

   c) Record all safety factors you need to consider.

3. With the approval of your teacher, demonstrate your model to the class.

(handwritten notes in margin:)
Say they need to demo — plate boundary operations to 1-3 graders. what could they use?

Materials
food (creme cheese)
snacks (candy bars) chocolate cake
wax
Silly Putty
Plasticine
clay
ice
shoe box
Styrofoam or foam pad

(handwritten note at bottom:)
?? Do mts form @ transform plate boundaries? Do valleys form.

Explore Questions

Design Investigation

Conduct Investigation

P 34 Investigating Earth Systems

Introducing Part B of the Investigation

See NSTA Pj Earth Science Geology
Activity 3 (Milky Way Bar)
or Readers Digest Book. See IES web
site (John Lahr shoe box model)

It should not be difficult for your students to devise more sophisticated ways of modeling subduction. One such way that is similar to the investigation in this activity but somewhat more realistic is to use a long strip of shower-curtain plastic, three or four inches wide and a few feet long, as the plate "conveyor belt" and a variety of readily obtainable thick but deformable materials, like cheese spread, peanut butter, or liquid margarine, to simulate the layer of deep-ocean sediment on the plastic strip. To model a more realistic subduction zone, two pieces of 2 x 4 lumber can be cut and arranged as shown in the sketch below. The plastic strip is then pulled down through the "subduction zone", and the deformable "sediment" is scraped off to form a strongly deformed mass, called an accretionary wedge, where the downgoing plate is subducted. The deformable layer might also be stratified, by use of thin layers of material with different colors, so that after the model is operated, the accretionary wedge could be sectioned with a kitchen knife (or, better, a cheese-cutting wire or taut fishing line) to reveal the deformation of the "sediments".

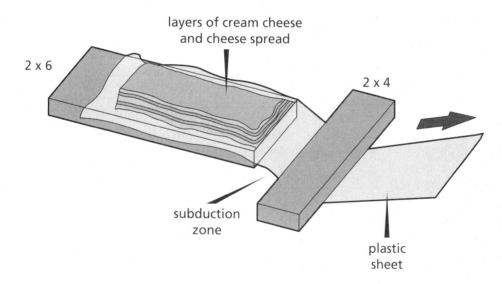

layers of cream cheese
and cheese spread

2 x 6

2 x 4

subduction
zone

plastic
sheet

Modeling sea-floor spreading is not as simple as modeling a subduction zone. Early in the plate-tectonics era there was an elegant and revealing experiment on the nature of spreading ridges. A shallow, tabletop-scale rectangular pan filled with melted wax was heated from below with heating elements and cooled from above by a fan. Conditions were adjusted so that a skim of solid wax formed at the surface. At two opposite edges of the pan, counter-rotating rollers pulled the plate of wax apart, in tension. A fracture formed across the wax plate, and as the two plates were pulled apart by the rollers, new "lithosphere" was formed at the "ridge crest". A number of other significant features of sea-floor spreading were reproduced. This is a beautiful example of how small-scale models can guide scientists' thinking. Such a model is beyond the resources of your students, but some might try to devise something simpler but similar in its purpose. If you have access to a library that has the journal called *Science* (a publication of the American Association for the Advancement of Science) or can obtain copies from another library, you can read the original paper in which this model was described. The full reference is cited below:

Oldenburg, Douglas W., and Brune, James N. (1972). *Ridge Transform Fault Spreading Pattern in Freezing Wax.* <u>Science</u>, v. 178, pages 301-304.

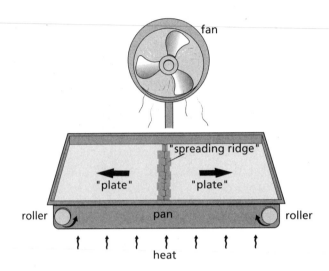

1. This part of the investigation encourages students to develop models for other kinds of plate boundaries. Students can model a divergent boundary or a transform boundary. Divergent boundaries were investigated in **Investigation 3**, and are briefly reviewed in **Digging Deeper** on page P36 of the Student Book. Transform boundaries are described on pages P36 and P37. Have students read the **Digging Deeper** reading section. Encourage them to also review the illustrations (which show the relative thicknesses and motions of lithospheric plates) and to make their models as realistic as possible.

    a) Students should record the type of plate boundary that they wish to model and develop a question to investigate. For example, students might suggest that they wish to study a transform boundary in order to investigate the question "Do mountains form at transform plate boundaries?"

    b) Check to make sure that students record their predictions. Following the example above, students might predict that "mountains will not form along a transform boundary because the plates will be moving parallel to one another, rather than against one another. The plates will slide against one another, but material will not be pushed upward to form mountains or downward to produce valleys."

2. One option is to provide a set of appropriate materials to students and suggest that they develop models from the available materials. Another option is to ask

students to construct the models at home (with adult supervision) using simple and inexpensive materials commonly available at home. Stress to students that simple materials can often be used to produce elegant models.

a) Students should note why they have selected materials for their models. To the extent possible, models should simulate the properties and behaviors expected. For example, if the students are modeling the flow of material, soft, deformable materials need to be used for the model to work properly rather than brittle or rigid materials.

b) Encourage students to write down their models clearly enough for another group to produce the model without additional instructions.

c) Stress that models must be safe. Provide some guidance to students about safety precautions, particularly with regard to using eye protection. Discourage students from designing a model that is intended to break or snap into pieces or require excessive force in order to operate.

d) Practice it

Teaching Tip

You may suggest that students do additional research to supplement their findings. From the standpoint of scientific inquiry, this is very important. Students need to understand that scientists do not simply engage in experiments. They also spend a great deal of time searching for relevant information to inform their investigations.

3. This is an excellent opportunity to reward creativity and ingenuity in design. Be sure that students focus their designs and presentations on the scientific process.

Assessment Tools

Investigation Journal Entry-Evaluation Sheet
Use this sheet to help students learn the basic expectations for journal entries that feature the write-up of investigations. It provides a variety of criteria that both you and your students can use to ensure that their work meets the highest possible standards and expectations. Adapt this sheet so that it is appropriate for your classroom, or modify the sheet to suit a particular investigation.

Group Participation Evaluation Forms I and II
Use these forms to provide students with an opportunity to assess group participation. Do not use results of this evaluation as the sole source of assessment data. Rather, it is better to assign a weight to the results of this evaluation and factor it in with other sources of assessment data.

Student Presentation Evaluation Form
Use the **Student Presentation Evaluation Form** as a simple guideline for assessing presentations. Adapt and modify the evaluation form to suit your needs. Provide the form to your students and discuss the assessment criteria before they begin their work.

Investigation 4: The Movement of the Earth's Lithospheric Plates

Digging Deeper

The Earth's Lithosphere

In the previous activity you learned that the rock of the Earth's mantle flows slowly in gigantic convection cells. The uppermost part of the mantle, however, does not take part in the convection. That's because its rock is not as hot, and it remains rigid while the rest of the mantle flows. Here's a similar example, on a much smaller scale. If you squeeze Silly Putty® at room temperature, it flows as you squeeze it in your hand. If you cool it in the refrigerator, it stays hard and rigid when you try to squeeze it. In **Investigation 3**, you found that this outermost rigid part of the Earth is called the lithosphere. The thickness of the lithosphere varies from place to place, but mostly it is a hundred or so kilometers. That's still fairly thin, compared to the thickness of the whole mantle, which is about 3000 km. The lithosphere has two parts: the Earth's crust, and the uppermost part of the mantle. The material below the lithosphere is called the asthenosphere ("weak sphere"). Unlike the lithosphere, the asthenosphere does take part in the convection of the mantle. The boundary between the lithosphere and the asthenosphere is really a temperature boundary. Below the boundary the rocks are hot enough to flow. Above the boundary they are cooler and rigid.

In the ocean basins the uppermost part of the lithosphere consists of the basalt that is formed by volcanoes along the mid-ocean ridges. This material is called the oceanic crust. It's only 4 to 8 km thick. The oceanic lithosphere gradually thickens as it moves away from the hot mid-ocean ridge. This is because the temperature boundary where the lithosphere turns into asthenosphere gets deeper in the Earth (see the diagram on the following page).

The Earth's continents form another part of the crust. The continental crust is very different from the oceanic crust.

As You Read...
Think about:

1. *What is the difference between crust and lithosphere?*
2. *What is the difference between oceanic crust and continental crust?*
3. *What is the difference between a subduction zone and a continent–continent collision zone?*
4. *Why do continents not go down subduction zones?*

Digging Deeper

This section provides text, illustrations, and a photo that give students greater insight into the nature of the lithosphere, lithospheric plates, subduction, and continent–continent collisions. You may wish to assign the **As You Read** questions as homework to help students focus on the major ideas in the text.

As You Read...

Think about:

1. The crust is the outermost shell of the Earth. The lithosphere consists of the crust and the upper, rigid part of the mantle (the part that does not take part in the mantle convection).

2. Continental crust is thicker, less dense, and generally much older than oceanic crust.

3. A subduction zone occurs where oceanic lithosphere is subducted under continental lithosphere (or where one oceanic lithospheric plate is subducted under another oceanic lithospheric plate). A trench forms on the sea floor as one lithospheric plate descends below the other. In a continent–continent collision, one of the two continental plates slides horizontally under the other for some distance. That thickens the continental crust and forms high mountain ranges.

4. Continents cannot go down subduction zones because continental lithosphere is less dense than the mantle.

Assessment Opportunity

You may wish to rephrase selected questions from the **As You Read** section into multiple choice or "true/false" format to use as a quiz. Use this quiz to assess student understanding and as a motivational tool to ensure that students complete the reading assignment and comprehend the main ideas.

It's thicker (mostly 30 to 50 km), its rock is less dense, and it's mostly very much older than the oceanic crust.

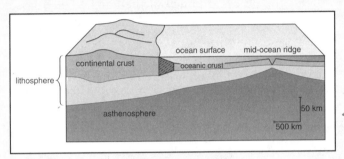

Lithospheric Plates

The lithosphere is not one continuous piece. Instead, it's made up of several very large pieces and a lot of smaller pieces. These pieces are called lithospheric plates (or just plates, for short). They fit together a bit like the pieces of a jigsaw puzzle. The line on the Earth's surface where two plates are in contact with each other is called a plate boundary.

Everywhere on Earth, the plates are in motion relative to one another. Along some boundaries, called divergent boundaries, plates are moving away from each other. Along other boundaries, called convergent boundaries, plates are moving toward one another.

The mid-ocean ridges, which you learned about in the last investigation, are divergent plate boundaries. As the plates move away from each other, new plate material is produced on either side of the ridge.

There is also a third kind of boundary, called a transform boundary, where the plates are moving neither towards one another nor away from one another. Instead, they are simply moving past one another like two cars in different lanes on the highway (only much slower!).

## Teacher Commentary

About the Photo

This is a generalized and simplified block diagram that shows the structure of the upper part of the Earth's interior. The difference in the horizontal and vertical scale bars shows that there is a lot of vertical exaggeration in the diagram. Although not labeled in the diagram, both the orange material (the asthenosphere) and the light gray layer above are part of the upper mantle. Note how the lithosphere (the crust and the upper mantle) thickens away from the mid-ocean ridge, because of the gradual cooling of the upper mantle as the material moves away from the mid-ocean ridge. It is likely to be confusing to your students that the uppermost part of the mantle is part of the lithosphere, and that the boundary between the lithosphere and the asthenosphere is simply a matter of the temperature of the mantle material. The boundary between the lithosphere and the asthenosphere becomes deeper just because the mantle cools on its journey away from the mid-ocean ridge. The entire lithosphere, from the mid-ocean ridge on the right to the left-hand edge of the diagram, is a single lithospheric plate, and there is no subduction at the contact between the oceanic crust and the continental crust. A continental margin of this kind is called a passive margin. The east coast of North America is a good example of a passive margin.

Sutured.

Teaching Tip

You may wish to use the following **Blackline Masters** to make overheads to use when discussing the **Digging Deeper** reading section:
- *Our Dynamic Planet* 4.2, Relative Motions of Major Lithospheric Plates,
- *Our Dynamic Planet* 4.3 Lithosphere,
- *Our Dynamic Planet* 4.4 Types of Plate Convergence,
- *Our Dynamic Planet* 4.5 Cross Section of Continent–Continent Convergence and
- *Our Dynamic Planet* 4.6 Collision of the Indian and Asian Plates.

An example of a transform boundary is the San Andreas Fault in California. It appears as a line in the aerial photo to the right. There, the Pacific Plate is moving northwest relative to the North American Plate. Lithosphere is neither created nor destroyed at transform boundaries. For this reason, transform boundaries are sometimes called conservative.

Subduction

Scientists have determined that the surface area of the Earth is not changing over time. Therefore, there must be plate boundaries where plates are consumed, as well as plate boundaries where plates are created. Plate boundaries where one plate dives down underneath another are called subduction zones. The downgoing plate consists of oceanic lithosphere. The other plate, the one that stays at the surface, can also consist of oceanic lithosphere, or it can be a continent. The place where the downgoing plate bends downward is marked by a deep trench on the ocean floor. Earthquakes and volcanoes are very common along subduction zones. The downgoing plate is eventually absorbed into the mantle, but scientists are just beginning to understand how that happens.

oceanic-continental convergence

oceanic-oceanic convergence

About the Photo

This aerial photo (note the blue wing of the airplane in the bottom of the photo) shows clearly the long linear scar on the Earth's surface in California known as the San Andreas Fault. The San Andreas Fault is a transform plate boundary, where the two plates slide parallel to one another with limited vertical motion along the boundary. The lateral (horizontal) offset on the fault is shown clearly by the offset of an active stream. The total movement on the fault is far greater than just the offset shown here, because, as the motion continues, the stream is offset so far that a new stream channel is finally formed on the downstream side of the fault, and the old channel is abandoned. This fault is said to be a right-lateral fault, because, whichever side you are standing on, if you step across the fault you seem to be displaced to the right. You might have your students pretend to do that. A fault that moves in the opposite sense is called a left-lateral fault.

About the Illustration

Ocean–ocean subduction and ocean–continent subduction are very similar, but there are some very significant differences. One of these is that in ocean–ocean subduction, the rising magma that is generated at a certain depth down the subduction zone rises up through mantle lithosphere, whereas in ocean–continent subduction, it rises through thick continental crust. Its passage through the continental crust contaminates it in such a way that the volcanoes are usually more explosive and dangerous. You might also point out to your students that these two diagrams represent "snapshots" of the subduction zones at a particular time. As time goes on, the subduction zones evolve, by receiving more and more sediment that is delivered to the subduction zone as it rides on the downgoing plate, and by increasing the volume of volcanic rock formed along the volcanic arc. In these and other ways, the subduction zones become more voluminous and more complex, geologically, with time.

Continent–Continent Collision

Subduction zones can make an ocean basin close up completely. When that happens, two continents meet at the subduction zone. Continents are less dense than the mantle, so they do not go down the subduction zone. It is like pushing a wooden board down into the water: the board tries to float up to the surface again. When the two continents meet, one of the continents is pushed horizontally beneath the other continent. The movement eventually stops, when the force of friction between the continents becomes large enough.

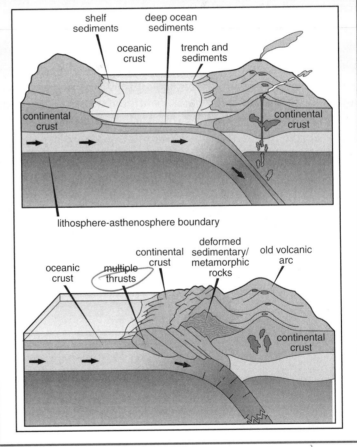

appalacians
(show maps)

About the Illustration

In the upper block diagram, a plate with both oceanic crust and continental crust is approaching a subduction zone where oceanic-crust lithosphere is being subducted under continental-crust lithosphere. There must have once been a mid-ocean ridge in between the two continents, but by now it has been "swallowed" down the subduction zone, thereby ceasing to exist. That can happen if the rate of production of new lithosphere at the mid-ocean ridge is less than the rate of consumption of old lithosphere at the subduction zone. The inevitable consequence is a continent–continent collision. The scene after the collision is shown in the lower block diagram. The structure of the newly combined continent at and near the line of collision (called the suture) is only now being worked out, mainly by studies of the geology and geophysics of the Himalayan region in southern Asia, the only good example of an active continent–continent collision today.

Investigation 4

Continent–continent collision zones are places where continents are thickest. Where a continent is thicker, it extends deeper down in the mantle, and its surface stands higher above sea level. There is one place on the Earth today where continent–continent collision is happening: India is slowly being pushed under southern Asia. That collision has produced the Himalayas, which are the highest mountains on Earth, and the Tibetan Plateau, which is the highest large plateau on Earth.

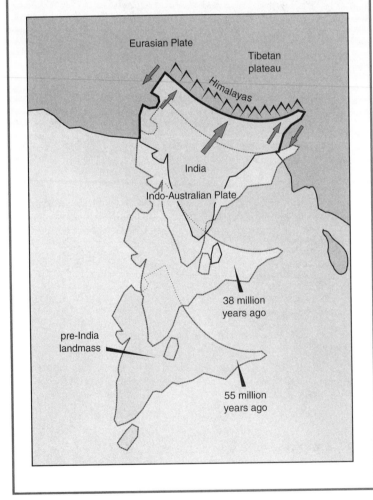

About the Illustration

This map is a "time-lapse" account of the approach of peninsular India, part of the Indo-Australian Plate, toward the Eurasian Plate, and their eventual collision. The first contact was millions of years ago, and the movement of India under southern Asia continues today. The underrunning, for hundreds of kilometers, has thickened the continental crust greatly. That thickening has created the Tibetan Plateau, the largest high plateau on Earth.

INVESTIGATING OUR DYNAMIC PLANET

Evidence
for Ideas

Review and Reflect

Review

1. What types of plate boundaries are found in or along the continental United States?

2. Why are folded mountain ranges found where plates converge? Why are folded mountains uncommon where plates move apart?

3. The Appalachian Mountains in the eastern United States are made up of folded rocks. Do you think that this suggests that this region was once the front edge of colliding plates, spreading plates, or sliding plates? Explain your reasoning.

Reflect

4. Why does the surface of a thicker continent stand higher above sea level than the surface of a thinner continent?

5. Would you expect volcanoes to form where plates slide past one another? Would you expect earthquakes? Explain your answers.

Thinking about the Earth System

6. Movement of the Earth's plates can create and destroy ocean basins. It can change the shape of oceans, the circulation of ocean water, and the depth of water in the ocean. Give an example of how plate tectonics might affect life on land or in the oceans (the biosphere).

7. When an oceanic plate is subducted beneath another plate, some water-rich sediment from the ocean crust descends into the mantle. There is also some water that is trapped within parts of the subducting plate that underlie the sediments. How does this process connect the hydrosphere and the geosphere?

Thinking about Scientific Inquiry

8. Why are models useful in scientific inquiry?

9. What are some problems associated with the use of models?

10. Describe how you used mathematics for science inquiry in this investigation.

Review and Reflect

Review

Allow your students ample time to pull all their evidence together and arrive at conclusions and explanations, and to make all the connections they can on the basis of their data.

1. There is a transform plate boundary along the western coast of the United States and a convergent plate boundary along the northwest coast of the United States. To provide greater detail, the transform boundary (the San Andreas Fault in California) extends northward from the Gulf of California and the convergent boundary occurs along northernmost California, Oregon, and Washington where the Juan de Fuca Plate is being subducted underneath the North American Plate. This has resulted in the formation of the Cascades Mountain Range. The eastern coast of the United States is not along a plate boundary, but is instead what is referred to as a passive continental margin. A continental margin is said to be passive where there is no relative motion between the part of the plate that carries continental crust and the part that carries oceanic crust. The eastern boundary of the North American Plate is the Mid-Atlantic Ridge, which is a divergent boundary.

2. Compressing rock together leads to folding. Folded mountains occur where plates converge because rocks at the leading edge of the plate are compressed (squeezed together). Folded mountains are uncommon where plates spread apart because the major forces are tensional forces rather than compression.

3. The Appalachian Mountains were probably at the leading edge of converging plates because folded mountains are common where plates converge.

Reflect

Allow students time to reflect on the nature of the evidence they have generated from their investigations. Again, help them see that evidence is crucial in scientific inquiry.

4. Continents are made of rock that is less dense than the mantle rock, so they float on the mantle. The thicker the continental crust, the higher it floats. It's like an ice cube in water: the thicker the ice cube, the more ice there is above water.

5. You would not expect volcanoes to form where plates slide past one another because sliding plates generally do not cause magma to form. You would expect earthquakes because, as the plates slide, stress builds up in the rocks, which is released suddenly to form earthquakes.

Thinking about the Earth System

6. If an ocean basin is created, it allows marine life to flourish. As an ocean basin is destroyed by subduction, the ocean gets smaller, which limits and eventually eliminates habitats for marine life. Continent–continent collision causes uplift of wide plateaus and the formation of high mountains, providing distinctive habitats for terrestrial plants and animals.

7. Some of the water trapped in sediments of the ocean crust, and also water that is incorporated into the minerals of the volcanic rocks while the rocks are still hot, are subducted into the mantle. When these rocks are heated, as they go deep down into the subduction zone, they give off the water, which is incorporated into magma. Volcanic eruptions return the water to the atmosphere and hydrosphere.

Thinking about Scientific Inquiry

Ask your students to share all the ways in which they communicated their findings to each other. Ask them if some of the ways that they used were more effective than others. Why was that? How could the class, as a whole, improve on its communication strategies?

8. Answers will vary. Students should note that in a matter of ten minutes in the lab they were able to model a process that takes millions of years to occur in nature (the collision of plates). This allowed them to better understand what happens where plates converge.

9. Models rarely replicate exactly what happens in nature because the conditions under which the model is run are different than the real-world event that the model represents. Models can help you to make predictions, but seldom can they replicate exactly what happens in nature.

10. Students will most likely note how they made measurements of the height and width of their models in **Part A** of the investigation.

> ### Assessment Tool
> **Review and Reflect Journal Entry-Evaluation Sheet**
> Use the general criteria on this evaluation sheet for assessing content and thoroughness of student work. Adapt and modify the sheet to meet your needs. Consider involving students in selecting and modifying the criteria for evaluating their reflections on **Investigation 4**.

NOTES

Teacher Review

Use this section to reflect on and review the investigation. Keep in mind that your notes here are likely to be especially helpful when you teach this investigation again. Questions listed here are examples only.

Student Achievement

What evidence do you have that all students have met the science content objectives?

_____

Are there any students who need more help in reaching these objectives? If so, how can you provide this?_____

_____

What evidence do you have that all students have demonstrated their understanding of the inquiry processes?_____

_____

Which of these inquiry objectives do your students need to improve upon in future investigations? _____

_____

What evidence do the journal entries contain about what your students learned from this investigation? _____

_____

Planning

How well did this investigation fit into your class time?_____

_____

What changes can you make to improve your planning next time? _____

_____

Guiding and Facilitating Learning

How well did you focus and support inquiry while interacting with students?

_____

What changes can you make to improve classroom management for the next investigation or the next time you teach this investigation? _____

_____

How successful were you in encouraging all students to participate fully in science learning? _____

_____

How did you encourage and model the skills values, and attitudes of scientific inquiry? _____

_____

How did you nurture collaboration among students? _____

_____

Materials and Resources

What challenges did you encounter obtaining or using materials and/or resources needed for the activity? _____

_____

What changes can you make to better obtain and better manage materials and resources next time? _____

_____

Student Evaluation

Describe how you evaluated student progress. What worked well? What needs to be improved? _____

_____

How will you adapt your evaluation methods for next time? _____

_____

Describe how you guided students in self-assessment. _____

_____

Self Evaluation

How would you rate your teaching of this investigation? _____

_____

What advice would you give to a colleague who is planning to teach this investigation? _____

_____

NOTES

INVESTIGATION 5: EARTHQUAKES, VOLCANOES, AND MOUNTAINS

Background Information

The field of study of plate motion is called plate tectonics. Tectonics comes from the Greek word *tekton*, which has to do with building or construction. Plate tectonics refers to the development of the features on Earth's surface due to deformation caused by plate movements.

In addition to the creation of mountain ranges, trenches, rift valleys, and ridges right at plate boundaries, other features and activity can result from plate movements. For example, there is a clear relationship between volcanoes, earthquakes, and plate boundaries. This is particularly evident around the rim of the Pacific Ocean, where the subduction of oceanic plates around much of the rim results in volcanic arcs and earthquakes.

The presence of a volcanic arc is good evidence for an active plate boundary. Volcanic arcs are supplied by magma that is generated deep in the mantle, above oceanic plates going down subduction zones. In order to melt asthenosphere to create magma, the downgoing plate must go deep enough to be surrounded by mantle rock at sufficiently high temperatures. The upper surface of the downgoing plate is covered with sediments that had been deposited on the plate during its long travel from a mid-ocean ridge to the subduction zone, plus additional sediment deposited in the trench from nearby land areas. Much of the sediment is plastered or accreted to the leading edge of the other plate, to build a growing and very complicated mass of sedimentary and metamorphic rock called an accretionary wedge (see below), but some of it is subducted deep down along with the downgoing plate. These sediments contain abundant water, and the rocks of the oceanic crust in the upper part of the downgoing plate also contain water, incorporated into the basaltic lavas in the rift valley of the mid-ocean ridge by the activity of hydrothermal springs, which are typically associated with the volcanism in the rift. At a sufficiently great depth down the subduction zone, much of the water is released and rises up into the overlying hot mantle asthenosphere. This addition of water lowers the melting point of the overlying rock, causing generation of magma by partial melting. The magma is slightly less dense than the solid rock from which it was derived, causing it to rise buoyantly toward the surface. When the magma reaches the surface, a volcano is formed.

A chain of such volcanoes behind a subduction zone forms a volcanic arc. Volcanic arcs are common in the Pacific Ocean along several ocean–ocean subduction zones, and they also extend along the Andes, run the length of Central America, across central Mexico, and run in a chain, known as the Cascades, from northern California to southern British Columbia. These latter volcanoes are associated with small oceanic plates (the Nazca, Cocos, and Juan de Fuca Plates, for example) that are being subducted under the Americas. The Alaskan Peninsula, the Aleutian Islands, the islands of Japan, the Philippines, and the Marianas are other examples of volcanic arcs.

Not all of the magma reaches the surface to form volcanoes; some of it remains below the surface in magma chambers and cools slowly

Investigation 5

to form massive and complex igneous intrusive bodies, called batholiths. Volcanic arcs thus typically consist of a foundation of intrusive igneous rocks overlain by thick successions of volcanic rocks. The Sierra Nevada, a major mountain range in central California, is thought to be mostly a gigantic and complex batholith that was the root of an ancient volcanic arc. Much later uplift, not associated with the subduction zone that formed the volcanic arc, has raised the batholith to the surface.

Not all volcanoes are associated with subducting plates. Hot spots are narrow plumes of unusually hot, and therefore less dense, mantle rock that is heated in certain places deep in the mantle, presumably at the interface between the mantle and the core. These plumes of rock then rise buoyantly toward the surface, flowing upward through the much broader convection cells of the asthenosphere. As they reach the upper mantle, where melting temperature is less because the pressure is much less, some of the hot rock melts, and the magma rises to the surface, resulting in hot-spot volcanism.

Hot spots can exist under continents as well as under oceans. The hot spot that has produced the hot springs and the past (and perhaps future!) volcanism at Yellowstone National Park is a good example. One theory suggests that the bulge created by a hot spot initiates the splitting of continents and the creation of rift (or divergent) zones. It is thought that a hot spot lies below, and is responsible for, the Rift Valley of Africa. There is also evidence suggesting that the New Madrid Fault, which runs down the Mississippi River Valley, may represent an aborted divergent zone originally created by a series of hot spots. The largest series of earthquakes in the United States outside of Alaska occurred on the New Madrid Fault in the early nineteenth century, ringing bells as far away as Philadelphia and causing the Mississippi River to run backward for a brief time! In this way, plate tectonics can affect even areas that are within the heart of a continent.

Hot spots tend to stay in one place relative to the deep mantle. As a plate moves relative to the deep mantle, the hot spot shifts its position on the moving plate. The Hawaiian Islands have been formed in just this way. The easternmost island, the "big island" of Hawaii, which has active volcanoes, is the youngest island in the chain, and is in the present position of the hot spot beneath. If you look in an atlas of the Pacific Ocean, you will see that there is a chain of seamounts (underwater mountains) called the Emperor Seamounts, which stretch to the northwest from the western end of the Hawaiian Islands chain. They represent even earlier positions of the hot spot relative to the Pacific Plate. The "dog leg" bend in the middle of the chain seems to represent a sudden change in the direction of movement of the Pacific plate long ago.

More Information...on the Web
Go to the *Investigating Earth Systems* web site www.agiweb.org/ies for links to a variety of other web sites that will help you deepen your understanding of content and prepare you to teach this module.

Investigation Overview

Students begin **Investigation 5** by considering the question: "How are earthquakes, volcanoes, and mountains related?" Students plot the locations of earthquakes, volcanoes, and major mountain ranges on a world map. They then search for patterns and relationships between these three events or features, and compare the locations of these features to a map of major plate boundaries. Then they conduct further research about these events and features and present their maps and findings to the class. Text in the **Digging Deeper** reading section explains the nature of earthquakes, the relationship between earthquakes and plate movement, the nature of volcanoes, the relationship between volcanoes and plate movements, the association of earthquakes and volcanoes, and the nature of mountain building.

Goals and Objectives

As a result of this investigation, students will develop a better understanding of the causes of earthquakes, volcanoes, and mountain building in relation to plate tectonics.

Science Content Objectives

Students will collect evidence that:
1. Earthquakes occur when rocks on either side of a fault slide past one another.
2. Volcanoes result from the eruption of molten rock, volcanic fragments, and gases at the Earth's surface.
3. Gas content often controls the explosiveness of a volcanic eruption.
4. Most earthquakes and volcanoes occur along plate boundaries.
5. Most of the world's major mountain chains are formed where two lithospheric plates collide.

Inquiry Process Skills

Students will:
1. Interpret data sets.
2. Plot coordinates on maps.
3. Look for patterns and relationships in data sets.
4. Draw conclusions about relationships between data sets.
5. Conduct additional research on earthquakes, volcanoes, and mountain chains.
6. Collate information into a useful format.
7. Communicate observations and findings to others.

Connections to Standards and Benchmarks

In this investigation, students investigate the causes of earthquakes, volcanoes, and mountain building. These observations will start them on the road to understanding the National Science Education Standards and AAAS Benchmarks shown on the following page.

NSES Links

- The solid Earth is layered with the lithosphere; hot, convecting mantle; and dense, metallic core.

- Lithospheric plates on the scales of continents and oceans consistently move at the rate of centimeters per year in response to movements in the mantle. Major geological events, such as earthquakes, volcanic eruptions, and mountain building, result from these plate motions.

- Landforms are the result of a combination of constructive and destructive forces. Constructive forces include crustal deformation, volcanic eruptions, and deposition of sediment, while destructive forces include weathering and erosion.

AAAS Links

- The interior of the Earth is hot. Heat flow and movement of material within the Earth cause earthquakes and volcanic eruptions and create mountains and ocean basins. Gas and dust from large volcanoes can change the atmosphere.

- Some changes in the Earth's surface are abrupt (such as earthquakes and volcanic eruptions) while other changes happen very slowly (such as uplift and wearing down of mountains). The Earth's surface is shaped in part by the motion of water and wind over very long times, which act to level mountain ranges.

Preparation and Materials Needed

Preparation

In **Steps 1 to 7** of this investigation, students plot the locations of earthquakes and volcanoes on a copy of a world map. They search for relationships between the locations of earthquakes, volcanoes, and major mountain chains. Make overhead transparencies of the **Blackline Masters** listed below. Use these overheads to refer to the maps and data while students are doing the investigation and during class discussion. Photocopy **Blackline Master** *Our Dynamic Planet* 5.1 World Map for each student. Have extra copies on hand for students who want to revise their work.

Blackline Master *Our Dynamic Planet* 5.1 World Map
Blackline Master *Our Dynamic Planet* 5.2 Table 1: Seismograph Station Results for Five Days
Blackline Master *Our Dynamic Planet* 5.3 Table 1: Global Volcanic Activity Over One-Month Period
Blackline Master *Our Dynamic Planet* 5.4 Major World Mountain Chains

Plot the data in **Tables 1 and 2** on a map to get a sense of how long it takes and to construct a key. You might also consider a second set of data on recent earthquakes and volcanoes for students to plot on the world map. You can obtain these data through the *Investigating Earth Systems* web site www.agiweb.org/ies. Most web-based data provide greater detail about latitude and longitude than is required for the scale of the map being used in the investigation. Keep the whole numbers for degrees of latitude and longitude, but remove the minutes and seconds before providing the data to students.

If you have a large world map that shows the names of countries (political world map), and/or a map that shows elevation and bathymetry (world geographic map), have these out for display during the investigation. **Tables 1 and 2** in the investigation provide geographic information about where the events occurred (e.g., Kuril Islands). Students can then consult the wall map to confirm that they plotted their data correctly. The geographic map will also reveal smaller mountain chains that do not appear on the map of major mountain chains on page P45.

In **Step 8** of the investigation, students research a particular volcano or earthquake and relate the event to the motions of Earth's lithospheric plates. To prepare for this research, gather resource materials and schedule access to computers for students to do research on the Internet. The *Investigating Earth Systems* web site www.agiweb.org/ies has links to helpful resources and web sites that will help students conduct research on earthquakes and volcanoes. Spend some time examining the links and resources at the site.

This investigation is closely connected to questions asked in the **Pre-Assessment**. The investigation will help students to develop better understandings of the causes of earthquakes and volcanoes, and to understand why and how earthquakes and volcanoes often occur together. To prepare for this investigation, review students' responses to the **Pre-Assessment** to consider their original ideas about these phenomena.

Materials

- colored pencils (3 different colors)
- copy of **Blackline Master** *Our Dynamic Planet* 5.1 World Map
- resource materials on volcanoes, earthquakes, and plate tectonics
- optional: Internet access for web-based research *

* The *Investigating Earth Systems* web site www.agiweb.org/ies has links to helpful resources and web sites that will help students conduct research on earthquakes and volcanoes.

NOTES

Investigation 5

Investigation 5:

Earthquakes, Volcanoes, and Mountains

Before you begin, first think about this key question.

How are earthquakes, volcanoes, and mountains related?

Explore Questions

In **Investigation 2** you discovered how earthquakes helped shed light on what is inside the Earth. You also found that earthquakes and volcanoes occur along subduction zones. Is that the only place that you find earthquakes and volcanoes? What makes them happen? How are mountains related to earthquakes and volcanoes?

Share your thinking with others in your group and with your class. Keep a record of the discussion in your journal.

Materials Needed

For this investigation your group will need:

• colored pencils
  (3 different colors)

• copy of a world map

Investigate

1. Discuss the following questions in your group. Be sure to explain your answers. Record the results of your discussion in your journal.

Key Question

Instruct students to respond to the **Key Question** in their journals. Allow a few minutes of writing time. Have them share their ideas about earthquakes, volcanoes, and mountains. Make a list of ideas on the chalkboard. Accept student ideas uncritically, even if they appear undeveloped (or are simply not correct). Do not praise wrong ideas, of course. The point of this exercise, as it is with all the **Key Questions**, is to provoke thought and prepare for the investigation.

Student Conceptions about the Relationships among Earthquakes, Volcanoes, and Mountains

Students commonly associate volcanoes with mountains. Many believe that volcanoes are the only kind of mountain, but some students appreciate that some mountains are volcanic, while others are caused by plates forcing the land up when plates collide. Students also associate earthquakes with volcanoes. Some think that volcanoes cause earthquakes, others think that earthquakes cause volcanoes, and still others hold both ideas. Some students in your class may believe that large earthquakes may form mountains. By the end of middle school, most students are familiar with the "Ring of Fire" and the presence of volcanoes along the Ring of Fire, but many students have less fully-developed ideas about why there are earthquakes along the Ring of Fire or how volcanoes and earthquakes can form in the middle of a plate. You can explore students' thinking about these ideas further by reviewing their answers to **Steps 1 (a) – (c)** of the investigation.

Answer for the Teacher Only

Earthquakes, volcanoes, and mountains are related, although only in part. Major earthquakes and volcanoes are associated with subduction zones. Mountains are associated as well with ocean–continent subduction zones. In continent–continent collision zones, earthquakes are common and mountains are formed, but volcanoes do not occur. It is not common for major mountain ranges not to be accompanied by earthquakes, at least at some stage in their development, but many mountain ranges are never accompanied by volcanic activity.

Assessment Tool

Key Question Evaluation Sheet

As with all assessment tools, let students know the basis for assessment at the start of an investigation. **The Key Question** is an assessment of students' prior knowledge and ideas. Students' preconceptions and beliefs *should not* be assessed for accuracy or science content. The criteria on the **Key Question Evaluation Sheet** include showing evidence of prior knowledge (explaining one's thinking and ideas) and reflecting discussion with classmates (revising or adding to work in light of ideas that emerge during discussion).

Investigation 5

About the Photo

The photo on page P41 looks north from the western edge of the Mid-Atlantic Ridge rift valley at Thingvellir, Iceland. The floor of the rift valley can be seen to the right of the photo, and the valley itself has formed from the combination of volcanic and tectonic forces (including earthquakes). Earthquakes and volcanoes are common occurrences in Iceland, where the North American and Eurasian plates meet. Iceland is a place where a mid-ocean ridge rises out of the ocean to form an island. This has happened because there the mid-ocean ridge lies atop a hot spot in the mantle. The result is that the ridge over this hot spot is shallower than elsewhere and is more volcanically active. These two factors have enabled the island of Iceland to rise out of the sea. Thingvellir is one of Iceland's most beautiful and historic places. It was here that the world's oldest existing parliament first met in AD 930.

Investigate

Teaching Suggestions and Sample Answers

Introducing the Investigation

Ask students to look at the questions **1a)** through **1c)**. Tell students that the investigation will give them an opportunity to check their ideas about where volcanoes and earthquakes occur. Provide students with necessary copies of handouts and colored pencils for **Steps 1 - 7** of the investigation. Middle-school students should be familiar with the meaning of latitude and longitude, but it may take them a few minutes to familiarize themselves again with these concepts. Resist the temptation to make this a teacher-led investigation by giving a lecture on latitude and longitude before students begin the investigation. As you visit student groups, check to see that they are working together. Check the first point in the data table on each student's map to make sure that the data are plotted correctly.

Blackline Master *Our Dynamic Planet* 5.1 World Map
Blackline Master *Our Dynamic Planet* 5.2 Table 1: Seismograph Station Results for Five Days
Blackline Master *Our Dynamic Planet* 5.3 Table 1: Global Volcanic Activity Over One-Month Period
Blackline Master *Our Dynamic Planet* 5.4 Major World Mountain Chains

1. The three questions that start the investigation are used to further explore student thinking about the causes of earthquakes and volcanoes and about students' ideas about the relationship between mountains and volcanoes. The idea explored in each question emerged from an examination of student work in the field test of the module.

How
Where do each form?

Always together

NOTES

Feature	Settings			Related feature & cause
Volc	arc	Cascades Italy ~~Japan~~	Indonesia	Eq, cone (out) shallow/deep
volc	MOR	Iceland		Eq yes (shallow) Mts yes
volc		Hawaii		Cone eq (shallow)
Mt	arc			
	Mor			
	Rockies appalachians			
EQ				

INVESTIGATING OUR DYNAMIC PLANET

Explore Questions

a) Can any mountain have a volcano erupt from it?

b) Do earthquakes and volcanoes always occur in the same area?

c) Do earthquakes and volcanoes always occur at the same time?

2. Look closely at *Table 1*. The data table shows recent earthquakes from around the world. The data was collected at regional seismograph stations.

Discuss the terms used in the table. Describe in your journal what the following terms mean and how they relate to the table.

a) Latitude   c) Depth

b) Longitude   d) Magnitude

Table 1: Subset of Seismograph Station Results for One Week				
Latitude	Longitude	Depth (kilometers)	Magnitude (Richter Scale)	Occurrence Region
47°N	151°E	141	5.2	Kuril Islands
28°S	178°W	155	5.0	Kermadec Islands
30°N	52°E	33	4.2	Iran
36°N	140°E	69	4.7	Honshu, Japan
34°N	103°E	33	4.3	Gansu, China
40°S	177°E	27	4.8	New Zealand
0°N	36°E	10	4.6	Kenya, Africa
38°N	21°E	33	4.6	Ionian Sea
16°N	47°W	10	4.7	N. Mid-Atlantic Ridge
6°S	147°E	100	4.4	New Guinea
55°N	164°W	150	4.5	Unimak Island, Alaska
24°S	67°W	176	4.1	Argentina
13°N	91°W	33	4.2	Guatemala coast
4°N	76°W	171	5.6	Colombia
40°N	125°W	2	4.5	N. California coast
5°S	102°E	33	4.4	S. Sumatra, Indonesia
44°S	16°W	10	4.6	S. Mid-Atlantic Ridge
51°N	179°E	33	4.4	Aleutian Islands
15°S	71°W	150	4.2	Peru
49°N	128°W	10	4.7	Vancouver, Canada
35°N	103°E	33	4.3	Gansu, China

Accept all reasonable responses to the questions. Answers for the teacher only are provided below. These questions can be answered as "yes" or "no" but encourage students to provide a reason for each answer. Tell students that the investigation will help them to test their ideas about earthquakes, volcanoes, and mountains.

a) The answer is definitely "no." Many mountains have no association with volcanic activity at any stage in their development. Most mountain ranges with volcanoes are those that are formed at subduction zones. Mountains that are formed in continent–continent collisions are usually not accompanied by volcanic activity. The mountains that are most likely to have a volcanic eruption are those that have been built by past volcanic activity.

b) "No," not always. Subduction zones are the characteristic places where both major earthquakes and major volcanic eruptions both occur. Along transform boundaries and in continent–continent collision zones, however, earthquakes are common but volcanic activity is uncommon or entirely absent. Both earthquakes and volcanoes are characteristic of mid-ocean ridges, but these are seldom near enough to populated areas to affect large numbers of people. Around islands produced by hot-spot volcanism, volcanoes are common but major earthquakes are not; Hawaii is a good example.

c) The answer to this is also definitely "no." Even in areas with both earthquakes and volcanoes, there are commonly major earthquakes that are not accompanied by volcanic eruptions, and there are commonly volcanic eruptions that are accompanied by only minor earthquakes. It is extremely uncommon for a major volcanic eruption and a major earthquake to happen at the same time.

2. Visit student groups. Check the first couple of data points against your map key to make sure that students have successfully worked together and understand the difference between latitude and longitude, and how to plot data points on a map.

a) Latitude is a measure of the north–south position of a point on the Earth relative to the Equator and the poles. It is measured as the angle between two lines: the line from the center of the Earth to a point on the Equator directly north (or south) of the given point, and the line from the center of the Earth to the given point itself. The latitude ranges from zero at the Equator to 90° at the poles.

b) Longitude is a measure of the east–west position of a point on the Earth. Lines of longitude, called meridians, extend due south from the North Pole, across the Equator, and to the South Pole. Conventionally, 24 equally spaced meridians are drawn on the Earth. Longitude is measured in degrees, 180° west and 180° east of what is called the Prime Meridian, which by convention passes through Greenwich, England.

c) Depth in the Earth is measured vertically downward from the Earth's surface toward the center of the Earth. Depth is usually expressed in kilometers. The place where an earthquake happens, somewhere in the Earth's interior, is called the focus (or hypocenter) of the earthquake. Depth to the earthquake focus can be as shallow as less than a kilometer or as deep as hundreds of kilometers.

d) Students will likely respond that magnitude describes the "strength," "power," "energy," or "force" of the earthquake. The magnitude of an earthquake is a measure of the energy released by the earthquake. There are several different scales in use for the earthquake magnitude. What is popularly called the Richter scale today is actually a modification of a scale originally developed by Charles F. Richter, a seismologist at the California Institute of Technology, in the 1930s to describe California earthquakes.

Teaching Tip

If you have a large world map that shows the names of countries and oceans, have this out for display. **Tables 1 and 2** in the **Investigation** provide geographic information about where the events occurred (e.g., the Kuril Islands). Students can then consult the wall map to confirm that they have plotted their data correctly.

Assessment Tools

Journal Entry-Checklist
Use this checklist as a guide for quickly checking the quality and completeness of journal entries.

Investigation Journal Entry-Evaluation Sheet
Use this sheet to help students learn the basic expectations for journal entries that feature the write-up of investigations. It provides a variety of criteria that both you and your students can use to ensure that their work meets the highest possible standards and expectations. Adapt this sheet so that it is appropriate for your classroom, or modify the sheet to suit a particular investigation. You should give the **Journal Entry-Evaluation Sheet** to students early in the module, discuss it with them, and use it to provide clear and prompt feedback.

NOTES

Investigation 5

Investigation 5: Earthquakes, Volcanoes, and Mountains

3. Use a copy of the world map shown.

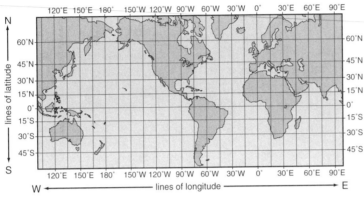

Conduct Investigations

a) Plot the locations of recent earthquakes shown in *Table 1*, using a colored dot for each earthquake.

b) Make a key at the bottom of the map to show what the dot represents.

Collect & Review

4. Use your copy of the map with the earthquakes plotted.

a) Plot the locations of recent volcanoes shown in *Table 2* on the next page. Use a different color and symbol than you did when plotting earthquakes.

b) Make sure your key reflects this new information.

P
43

Investigating Earth Systems

3. a) This should not be difficult for students if they understand the concept of interpolation. Longitude is labeled in 30° increments and latitude is marked in 15° increments, but the data are supplied to the nearest whole degree. You might need to help some students in placing the data points in the correct interpolated positions. Two interpolations are needed for each data point: one for latitude (north–south position) and the other for longitude (east–west position). This is done "by eye" and will necessarily be somewhat approximate.

   b) This can be as simple as a single dot, colored to correspond to the dots they plotted on the map, along with words like "earthquake location." To make better use of the data, however, you might help the students to develop a number of regularly spaced categories of earthquake magnitude, each with its own color. For example, red might represent an earthquake of magnitude 4.0 to 4.5, blue for magnitude 4.6 to 4.9, and so on. It would be good to use at least four categories. Some students might want to use many more, but it would be good to limit the number of categories to no more than six or eight; otherwise the multitude of colors becomes confusing, and the device loses some of its usefulness.

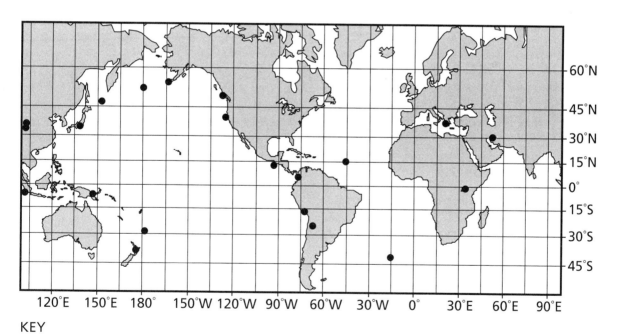

KEY

● Earthquakes

INVESTIGATING OUR DYNAMIC PLANET

Table 2: Global Volcanic Activity Over One-Month Period			
Latitude	Longitude	Location	Region
1°S	29°E	Nyamuragira	Congo, Eastern Africa
38°N	15°E	Stromboli	Aeolian Islands, Italy
37°N	15°E	Etna	Sicily, Italy
15°S	71°W	Sabancaya	Peru
0°	78°W	Guagua Pichincha	Ecuador
12°N	87°W	San Cristobal	Nicaragua
0°	91°W	Cerro Azul	Galapagos, Ecuador
19°N	103°W	Colima	Western Mexico
19°N	155°W	Kilauea	Hawaii, USA
56°N	161°E	Shiveluch	Kamchatka, Russia
54°N	159°E	Karymsky	Kamchatka, Russia
43°N	144°E	Akan	Hokkaido, Japan
39°N	141°E	Iwate	Honshu, Japan
42°N	140°E	Komaga-take	Hokkaido, Japan
1°S	101°E	Kerinci	Sumatra, Indonesia
4°S	145°E	Manam	Papua, New Guinea
5°S	148°E	Langila	Papua, New Guinea
15°S	167°E	Aoba	Vanuatu
16°N	62°W	Soufriere Hills	Montserrat, West Indies
12°N	86°W	Masaya	Nicaragua
37°N	25°W	Sete Cidades	Azores

Inquiry

**Using Maps and Data
Tables as Scientific Tools**

*Scientists collect and review
data using tools. You may think
of tools as only physical objects
such as shovels and hand
lenses, but forms in which
information is gathered, stored,
and presented are also tools
for scientists. In this
investigation you are using
scientific tools: data tables
and maps.*

5. Use your map to answer the following questions:

   a) List several locations where an earthquake happened
   close to a volcanic eruption.

   b) List several locations where an earthquake happened far
   from the nearest volcanic eruption.

   c) Describe any pattern or patterns in the locations of
   earthquakes and volcanoes.

   d) How might additional data help you to find patterns,
   trends, and relationships between volcanoes and
   earthquakes?

6. The next map shows the major mountain chains of
   the world.

Evidence
for Ideas

Consider
Evidence

4. a) The locations of the volcanoes can be plotted in the same way as the locations
        of the earthquakes but with symbols of a different color and perhaps also a
        different shape (triangles vs. circles, for example). If the earthquakes were
        plotted as several categories by magnitude, using different colors, then make
        sure that the symbols for the volcanoes are very different from those for the
        earthquakes, so that a viewer of the map can distinguish immediately between
        earthquakes and volcanoes.

    b) As you circulate to groups, check to make sure that students have made a key.
        If you see a map without a key, ask the student: "What does the green circle
        represent on your map?" This will help them to appreciate why a map key is
        useful—it helps another person to know what the student has done.

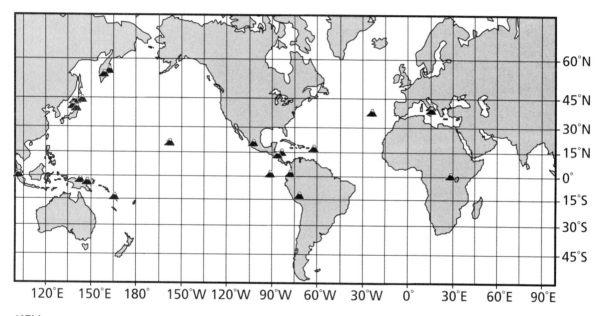

KEY

▲ Volcanoes

5. a) Answers will vary depending upon how the word "close" is interpreted, but
        there are several areas where an earthquake happened close to a volcanic
        eruption. Examples include New Guinea, Indonesia, and Japan.

    b) There are many such locations. The data sets represent a very small period
        of time (one week or one month). It should be clear to students that volcanoes
        and earthquakes do not always occur together, nor do they always occur at the
        same time.

    c) There appears to be more earthquakes and volcanoes around the rim of the
        Pacific Ocean (earthquakes and volcanoes commonly appear together along
        ocean–ocean and ocean–continent subduction zones). Some earthquakes
        appear in areas without volcanic eruptions, and vice versa.

Investigation 5

d) The more data points, the more clearly the spatial relationships between earthquakes and volcanoes will emerge. The data sets for both earthquakes and volcanoes given here are rather "thin."

6. If you have a large geographic world map, this would be a good time to display it in the classroom. The diagram on page P45 of the Student Book shows only major mountain chains. Having a more detailed map will be helpful to students because it will reveal other mountainous regions that have earthquakes and/or volcanoes.

NOTES

Investigation 5

Investigation 5: Earthquakes, Volcanoes, and Mountains

Collect & Review

a) Add this information to your map. Use a different color and a symbol to represent the mountains.

b) Make sure your key reflects this new information.

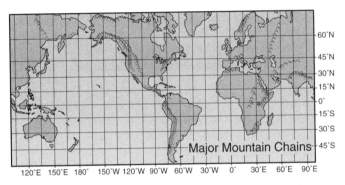

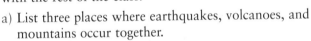Major Mountain Chains

7. You now have a map showing the correlation between earthquakes, volcanoes, and major mountain chains in the world. Compare your map to the map of plate boundaries on P27. Discuss and record your ideas about the following. Share your observations and discuss your ideas with the rest of the class.

Evidence for Ideas

a) List three places where earthquakes, volcanoes, and mountains occur together.

b) List three volcanic mountain chains.

c) Explain the relationship you think there is among earthquakes, volcanoes, and mountains.

Collect & Review

8. The *IES* web site lists several eathquakes and volcanoes. Each group should choose one volcano and one earthquake to investigate. You may wish to divide this task so that each group takes responsibility for a different information source. These might include:

- Earth science textbooks and reference books.
- Encyclopedias.
- CD-ROMs.
- The *IES* web site at www.agiweb.org/ies.

Show Evidence

Work together to use what you've learned about plate tectonics to explain why these earthquakes and volcanoes occurred. Discuss your findings with the rest of the class.

a) Record the results of your discussions in your journal.

Inquiry

Correlations as Evidence

A correlation is a relationship or connection between two or more things. Correlations are often the first kind of evidence gathered when trying to explain an occurrence.

P
45

a) Symbolizing the mountain ranges is not as simple as for earthquakes and volcanoes, because the mountain ranges are long belts, not points. Have your students think carefully about the best way to symbolize them.

b) As you circulate, look at student work to see if students have added a new symbol on the key for their maps.

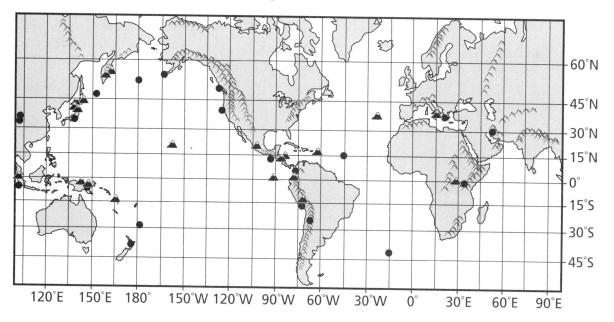

KEY

⤳ Mountain Ranges   ▲ Volcanoes   ● Earthquakes

7. As students work, refer them to **Investigation 4**, specifically the maps on page P27 and P33 of the student text. They will note that earthquakes and volcanoes commonly appear together along ocean–ocean and ocean–continent subduction zones.

a) Examples include: west coast of South America; Kamchatka, Russia; Central America. Certain other mountainous areas fall into this category but are not represented as "major mountain chains" on the map on page P45: for example, Japan, Indonesia, and New Guinea.

b) Andes; Sakhalin; Central America. (However, see comment under **Step 7(a)** above.)

c) See comments under **Step 1** on page 192.

Investigation 5

8. You will need to decide about the extent to which you wish to hold students accountable for explaining the plate tectonic setting. At the very least, students should be asked to identify whether or not the earthquake or volcano occurred at a plate boundary. Beyond that, you can ask that they specify the type of boundary at which the event occurred. The deeper you have students go in their explanations, the greater their understanding about the nature of the causes of earthquakes and volcanoes is likely to be.

a) Have students report briefly on their work in a class discussion. Alternatively, post the results of their work in the classroom and allow time for students to visit the work and to ask questions. Students should record the main ideas of their review of work in their journals.

NOTES

Investigation 5

INVESTIGATING OUR DYNAMIC PLANET

Digging   Deeper

Evidence
for Ideas

As You Read...
Think about:
1. *What is the cause of
   earthquakes?*
2. *How are faults and
   earthquakes related?*
3. *What is the cause of
   volcanoes?*
4. *How does gas content
   affect how a volcano
   erupts?*
5. *How is volcanism at a
   hotspot different
   from volcanism at a
   mid-ocean ridge. How
   are they similar?*
6. *Why are mountains
   found in regions where
   the lithosphere is thick?*

EARTHQUAKES, VOLCANOES, AND MOUNTAINS

The Nature of Earthquakes

Like all solids, rocks have strength. It takes a large force to break them. Plate movements cause large forces to build up within the lithosphere, and at certain times and places, the forces become greater than the strength of the rock. The rock then breaks, along a fracture surface that sometimes extend for tens of kilometers. This surface is called a fault. Faults are fractures in the Earth's surface along which there has been rupture and movement in the past. When the rocks break, the rocks on either side of the fracture plane slide past one another, until the forces are relieved. Strong vibrations are produced as the rock masses slide past one another. Those vibrations are felt as an earthquake. The vibrations travel away in all directions in the form of seismic waves, which you learned about earlier. Over time, the fracture "heals," making the strength of the rock greater again. For this reason, faults tend to slip, then stick, then slip again, and so on. As the rocks on either side of a fault slide past one another, they produce strong vibrations.

Earthquakes and Plate Movements

Many of the largest earthquakes occur along subduction zones, as the downgoing plate slides downward. The pattern of forces within the plate that cause the earthquake fracture is very complicated. The result of those forces, however, is the occurrence

Digging Deeper

This section provides text and photos that give students greater insight into the nature of earthquakes, the relationship between earthquakes and plate movement, the nature of volcanoes, and the relationship between volcanoes and plate movements, the association of earthquakes and volcanoes, and the nature of mountain building. You may wish to assign the **As You Read** questions as homework to help students focus on the major ideas in the text.

As You Read...

Think about:

1. Earthquakes are caused by movement along a fracture plane in the Earth's crust. When the rocks break, rocks on either side of the fracture plane slide past one another, and the strong vibrations that result are felt as an earthquake.

2. Faults are fractures in the Earth's crust along which there has been rupture and movement in the past. Earthquakes occur when the forces that build up along faults overcome the strength of the rocks, the rocks on either side of the fault slide past one another, and strong vibrations are generated.

3. Volcanoes are caused by the rise of magma formed within the Earth. Magma rises because it is less dense than surrounding rock. The rising magma does not always reach the surface, but when it does, it forms a volcano.

4. If the gas content of a magma is low, the magma erupts gently. If the gas content of the magma is high, an explosive and violent eruption can occur.

5. Volcanism at a hot spot can occur far from a plate boundary, whereas volcanism at a mid-ocean ridge occurs at a plate boundary (a divergent boundary). Both types of volcanism involve the production and release of magma.

6. When plates collide, the lithosphere becomes thicker. Because the lithosphere floats in the mantle beneath, the thicker the lithosphere, the higher its top will be. Like a block of wood floating in water, the height of wood that stands above the surface of the water depends upon the thickness of the wood.

About the Photo

Severe earthquakes can cause large buildings to collapse from the shaking of the ground beneath the buildings. In this particular case, the land upon which homes in the Marina district of San Francisco were built was actually fill from the great San Francisco earthquake of 1906. The fill shook violently during a major earthquake in October 1989 and reduced many buildings from three stories to one story in height in moments. Note the railing on the patio in the right side of the photograph—most likely a deck on the second or third floor of the building.

of earthquakes that range from very shallow to as deep as hundreds of kilometers. Earthquakes are also very common in continent–continent collision zones, as one continent is pushed beneath another.

Along transform boundaries, the two plates slide parallel to one another along a surface called a transform fault. If the fault becomes locked for a long period of time and then suddenly slips, a major earthquake results. Along some transform boundaries, however, the fault slips continuously, causing nothing more than very minor earthquakes. The San Andreas Fault, in California, is an unusually long transform fault. It is locked in southern California, in the vicinity of Los Angeles. It is also locked in central California, in the vicinity of San Francisco. That is why the earthquake hazard is great in both of those cities.

The Nature of Volcanoes

A volcano is a place where molten rock, and also solid volcanic fragments and volcanic gases, are erupted at the Earth's surface. At certain times and places, rock deep in the Earth is melted, to form magma. The magma rises upward, because it is less dense than the surrounding rock. It does not always reach the surface before it crystallizes to rock again, but when it does, it forms a volcano.

Volcanoes vary a lot in how they erupt. The most important factor is the gas content of the magma. All magmas have gases dissolved in them, in the same way that soft drinks have carbon dioxide dissolved in them to make them fizzy. As the magma gets close to the surface, the pressure on the magma decreases. That causes some of the gas to bubble out of the magma. Magma with low gas content comes out of the volcano without violent explosions and then flows peacefully down the sides of the volcano. Magmas with high gas content cause powerful explosions when they approach the surface.

→

Assessment Opportunity

After students have completed the **As You Read,** ask them to revisit their responses to pre-assessment to see what they wrote for the first two questions (these questions match **Questions 1** and **3** of **As You Read.** Ask them to explain why their answers have changed since the beginning of the module. This provides you with information about specific work within the module that has led to a change in student understanding.

Investigation 5

The explosions blow globs of magma and pieces of broken rock high into the atmosphere. Large explosive volcanic eruptions are the most serious hazard humankind faces, except for extremely rare impacts of large meteorites.

Volcanoes and Plate Movements

In **Investigation 3** you learned how volcanoes are formed along mid-ocean ridges. Those volcanoes are very numerous, but most of them are deep in the ocean. In some places, however, volcanic activity on a mid-ocean ridge is strong enough to build an island above sea level. Iceland is a good example of that.

Most large volcanoes occur along subduction zones. Certain scientists think that some volcanoes near subduction zones are caused when parts of the subducting ocean crust reach a certain depth and begin to melt. Many scientists, however, believe that the cause of most subduction zone volcanoes has to do with the water that is contained in the rocks of the ocean crust. At a certain depth down the subduction zone, the water is released from the rocks. The water rises up into the mantle above the subducted plate. Laboratory experiments have shown that adding water lowers the melting point of the mantle rocks. Whichever way the magma is generated, it rises up to feed volcanoes along the subduction zone.

There is another kind of volcano called a hot spot volcano. It is caused by a hot spot in the mantle that generates magma for long periods of time. Scientists think that hot spots don't move, and so a line of volcanoes forms as the plate moves over the hot spot. The orientation of that line and ages of the volcanoes that make it up reveal the direction and speed of plate movement. Unlike most other volcanoes, hot spot volcanoes can occur far from plate boundaries. The Hawaiian Islands and Yellowstone Park are good examples of hot spot volcanism.

NOTES

Investigation 5

Investigation 5: Earthquakes, Volcanoes, and Mountains

The Association of Earthquakes and Volcanoes

Along subduction zones, major earthquakes and large volcanoes are both common. Most of the Pacific Ocean is rimmed with subduction zones. That's why earthquakes and volcanoes are so common in countries that border the Pacific. You might have heard that the Pacific Rim is called the "Ring of Fire." In continent–continent collision zones, as in southern Asia, earthquakes are common but volcanoes are not formed. Countries like China, India, Iran, and Turkey experience major earthquakes but not volcanoes.

Mountain Building

Most of the world's large mountain ranges are formed where two lithospheric plates collide. Where two plates converge at a subduction zone, enormous volumes of material are added to the region. Some of this material is sediment that is scraped off from the downgoing plate. Also, magma from deep in the subduction zone rises up to feed volcanoes on the plate that isn't subducting. Some of the magma stays below the surface to form deep igneous rocks. As the crust near the subduction zone grows in volume, its base becomes lower and its top becomes higher. It's very much like a block of wood floating in water: the thicker the block, the lower its base, and the higher its top. The rocks of the Earth's lithosphere float on the denser mantle below, so when the lithosphere becomes thicker, mountains are formed. The Andes, along the west coast of South America, have been formed in that way. The same thing happens when two continents collide. As one of the continents is shoved beneath the other, the lithosphere becomes thicker, so it rises up to form a mountain range. The Himalayas, in southern Asia, have been formed in that way.

About the Photo

Ask students what they think caused the mountain in the photograph. Students commonly associate mountains with volcanoes. The shape of the mountain in the photograph may lead students to assign a volcanic origin to the mountain. Mountains can form in other ways, as when the crust is folded and compressed and lithosphere is thickened when plates collide. The peaks of these mountains can be shaped through glacial erosion into forms very similar to volcanic mountains. Some of the world's highest mountains, including the mountains shown in this photograph, are not volcanic.

INVESTIGATING OUR DYNAMIC PLANET

Evidence for Ideas

Review and Reflect

Review

1. Review your answers to **Investigate, Step 1**. Answer the questions again, using what you learned in this investigation. Be sure to explain your answers.

   a) Can any mountain have a volcano erupt from it?

   b) Do earthquakes and volcanoes always occur in the same area?

   c) Do earthquakes and volcanoes always occur at the same time?

2. Where do most earthquakes occur in the United States? Why?

3. Where in the United States are most volcanoes found? Why?

Reflect

4. How does the gas content of a magma affect the shape of a volcano?

5. Dynamic means powerful or active. How has this investigation added to your understanding of Earth as a dynamic planet?

Thinking about the Earth System

6. How does the hydrosphere influence the nature of volcanic eruptions?

7. How do volcanic eruptions affect the atmosphere?

8. How are volcanoes, earthquakes, and mountains (geosphere) linked to the biosphere?

   Remember to write in any connections that you have made between volcanoes, earthquakes, mountains, and the Earth System on your *Earth System Connection* sheet.

Thinking about Scientific Inquiry

9. When did you form your hypotheses in this investigation?

10. What scientific tools did you use in this investigation?

Review and Reflect

Review

Allow your students ample time to pull all their evidence together and arrive at conclusions and explanations. Help them make all the connections based upon their data.

1. a) No. Some mountains form when the crust is folded and compressed and lithosphere is thickened when plates collide.

    b) No. Earthquakes and volcanoes do not always occur together. In transform plate boundaries like along the San Andreas Fault in southern California, earthquakes occur as a result of faulting, but magma that is needed for volcanoes to form is not produced along such zones. Another example of earthquakes without volcanism is the continent–continent collision zone in southern Asia.

    c) No. Earthquakes commonly occur during the movement of magma during or preceding a volcanic eruption, but earthquakes can occur without volcanism.

2. Most earthquakes in the United States occur along faults that are related to the transform boundary that separates the North American Plate from the Pacific Plate in California.

3. Most volcanoes in the United States occur in the northwestern United States, where oceanic lithosphere is being subducted beneath continental lithosphere.

Reflect

Give students the opportunity to link the scientific part of their investigation to social considerations. It is important that they begin to appreciate that human activity can often result in changes to the natural environment. It is also important that students recognize how humans have adapted the natural environment to solve their needs, and that sometimes this can cause conflicts.

4. When a volcano has high gas content, the volcano is likely to erupt explosively, blasting fragments of volcanic rock and ash into the air, followed by ashflows down the slopes of the volcano. The volcano can build a large cone of rock and ash. When the gas content of the magma is low, magma flows gently down the sides of the volcano, forming broad volcanic mountains.

5. Answers will vary. Encourage students to think about the driving forces between the process of mantle convection and the resulting formation of volcanoes and earthquakes at the surface.

Investigation 5

Thinking about the Earth System

Investigation 5 will have enhanced your students' knowledge of the relationships between earthquakes, volcanoes, and mountains. They now need to reflect on this in terms of the Earth System. Help them to connect what they have learned with the geosphere, the atmosphere, the hydrosphere, and the biosphere.

6. The connection between the hydrosphere and volcanism was discussed on page P48 of the **Digging Deeper** reading section. In many places, groundwater is incorporated into magma when it reaches shallow levels in the crust. The added water makes a volcanic eruption more likely to be explosive, and therefore dangerous.

7. Major volcanic eruptions put fine volcanic ash high in the atmosphere, where it spreads worldwide and takes months or even years to settle back to the Earth's surface. In the meantime, it blocks some of the sunlight reaching the Earth, causing lower-than-normal temperatures around the globe. Gases like sulfur dioxide that are emitted by volcanoes also have a temporary cooling effect on the Earth's climate.

8. Volcanic eruptions can kill all living things over a large area around the volcano. Earthquakes have little direct effect on the biosphere, except to kill some animals (including humans!) in the immediate vicinity. The development of mountain ranges can cause great changes in the distribution of plants and animals, by changing the climate both upwind and downwind of the mountain range.

Teaching Tip
Remind students to enter any new connections that they have found on the *Earth System Connection* sheet.

Thinking about Scientific Inquiry

Give students time to reflect on the nature and limitations of the models they used. Again, help them see that modeling is useful to scientific inquiry.

9. When developing ideas about the interrelationships among earthquakes, volcanoes, and mountains.

10. Plotting different kinds of data, by interpolation, on a single map.

Assessment Tool

Review and Reflect Journal Entry-Evaluation Sheet

Use the general criteria on this evaluation sheet for assessing content and thoroughness of student work. Adapt and modify the sheet to meet your needs. Consider involving students in selecting and modifying the criteria for evaluating their reflections on **Investigation 5**.

Teacher Review

Use this section to reflect on and review the investigation. Keep in mind that your notes here are likely to be especially helpful when you teach this investigation again. Questions listed here are examples only.

Student Achievement

What evidence do you have that all students have met the science content objectives?

_____

Are there any students who need more help in reaching these objectives? If so, how can you provide this?_____

_____

What evidence do you have that all students have demonstrated their understanding of the inquiry processes?_____

_____

Which of these inquiry objectives do your students need to improve upon in future investigations? _____

_____

What evidence do the journal entries contain about what your students learned from this investigation? _____

_____

Planning

How well did this investigation fit into your class time?_____

_____

What changes can you make to improve your planning next time? _____

_____

Guiding and Facilitating Learning

How well did you focus and support inquiry while interacting with students?

_____

What changes can you make to improve classroom management for the next investigation or the next time you teach this investigation? _____

_____

How successful were you in encouraging all students to participate fully in science learning? _____

_____

How did you encourage and model the skills values, and attitudes of scientific inquiry? _____

_____

How did you nurture collaboration among students? _____

_____

Materials and Resources

What challenges did you encounter obtaining or using materials and/or resources needed for the activity? _____

_____

What changes can you make to better obtain and better manage materials and resources next time? _____

_____

Student Evaluation

Describe how you evaluated student progress. What worked well? What needs to be improved? _____

_____

How will you adapt your evaluation methods for next time? _____

_____

Describe how you guided students in self-assessment. _____

_____

Self Evaluation

How would you rate your teaching of this investigation? _____

_____

What advice would you give to a colleague who is planning to teach this investigation? _____

_____

Investigation 5

INVESTIGATION 6: EARTH'S MOVING CONTINENTS

Background Information

The forerunner to the theory of plate tectonics was first proposed as the theory of continental drift early in the twentieth century by Alfred Wegener, a German natural scientist. As evidence for his theory, Wegener cited the modern distribution of several distinctive kinds of mineral deposits, rock formations, and fossils, which are now widely separated but which must have been together earlier in geologic time. These features were especially useful in implying that South America and Africa were once joined. The consistently good match of such features among all of the southern hemisphere continents convinced many southern hemisphere geologists in the first half of the twentieth century that all of the continents were once joined together in the form of a single continent, which geoscientists now call a supercontinent. The theory was not widely accepted among northern hemisphere geologists, however, in part because the evidence for the existence of a supercontinent was not quite as strong for the northern hemisphere continents but also because geophysicists saw no mechanism by which the continents could move relative to one another. Soon after mid-century, however, the geophysicists turned the tables on themselves, by using evidence from the Earth's magnetic field that showed convincingly that the continents have indeed moved relative to one another and relative to the deep interior of the Earth. By 1970, the reality of continental drift was almost universally accepted, in the form of the theory of plate tectonics.

The assembly of the supercontinent of Pangea involved several continent–continent collisions along continental boundaries, when earlier oceans were finally closed by continuing subduction. The result was a number of long, sinuous belts, called orogenic belts, in which the geological record shows abundant evidence of uplift to form high mountain chains and plateaus, as well as deformation, metamorphism, and igneous activity. The Himalayas are the modern example of such an orogenic belt.

Eventually, probably because the insulating effect of the thick Pangea continent caused a gradual rise in temperature in the underlying mantle, Pangea was rifted apart into several pieces. Rifts develop when heating from below bows the lithosphere upward. The reason for the upward bowing is not a force from below; rather, the heating causes upward expansion. The two sides of the rift then slowly move apart, down the slope caused by the increased elevation. As the gap between the continents widens, the ocean invades the rift. The ocean then gradually increases in width. The Atlantic Ocean is a good modern example of an ocean that has formed in this way. The Atlantic Ocean is widening by several centimeters per year. Just in recent years, the development of GPS (the satellite-based global positioning system) has allowed direct measurement of this movement, on a year-to-year basis.

More Information...on the Web
Go to the *Investigating Earth Systems* web site www.agiweb.org/ies for links to a variety of other web sites that will help you deepen your understanding of content and prepare you to teach this module.

Investigation Overview

At the beginning of this investigation, students are asked whether or not they think that continents and oceans have always been in the same place that they are today. This question challenges them to think about the "permanence" of continents and ocean basins. In the investigation, students explore the evidence needed to evaluate whether or not continents and oceans have moved over time. Students examine the fit of the continents across the Atlantic Ocean, the patterns of mountain belts, evidence from fossils, and evidence from glaciers to reconstruct what the Earth's continents and oceans looked like 250 million years ago. The **Digging Deeper** reading section reviews the theory of continental drift, the concept of supercontinents (including Pangea), and the breakup of Pangea. Thus, the investigation challenges students to relate the events that they have studied (the causes and locations of earthquakes, volcanoes, and mountains) to evidence for a scientific theory that explains the occurrence of these events and features of the Earth.

Goals and Objectives

As a result of this investigation, students will understand how evidence supports the idea that the positions of continents and ocean basins change over geologic time.

Science Content Objectives

Students will collect evidence that:
1. Continent shapes appear to fit together.
2. There are common fossils, mountain chains, and glacial deposits on different continents, some of which are now widely separated by oceans.
3. The theory of continental drift fell out of favor with scientists until a mechanism (plate tectonics) for moving the continents was discovered.

Inquiry Process Skills

Students will:
1. Collect evidence from maps and other sources of information to support or refute a theory.
2. Analyze data, looking for patterns and relationships.
3. Use data analyses to support or refute a theory.
4. Revise conclusions based upon new evidence.
5. Share findings and conclusions with others.

Connections to Standards and Benchmarks

In **Investigation 6**, students learn about the evidence for continental drift. These observations will start them on the road to understanding the National Science Education Standards and AAAS Benchmarks shown on the following page, along with other standards and benchmarks related to the history and nature of science.

NSES Links

• Lithospheric plates on the scales of continents and oceans consistently move at the rate of centimeters per year in response to movements in the mantle. Major geological events, such as earthquakes, volcanic eruptions, and mountain building, result from these plate motions.

• The Earth processes we see today, including erosion, movement of the lithospheric plates, and changes in the atmospheric composition, are similar to those that occurred in the past.

• Fossils provide important evidence of how life and environmental conditions have changed.

• Evidence consists of observations and data on which to base scientific explanations.

AAAS Links

• Thousands of layers of sedimentary rock confirm the long history of the changing surface of the Earth and the changing life forms whose remains are found in successive layers. The youngest layers are not always found on top, because of folding, breaking, and uplift of layers.

• Some changes in the Earth's surface are abrupt (such as earthquakes and volcanic eruptions) while other changes happen very slowly (such as uplift and wearing down of mountains). The Earth's surface is shaped in part by the motion of water and wind over very long times, which act to level mountain ranges.

Investigation 6

Preparation and Materials Needed

Preparation

You should try this investigation before teaching it. This will give you a sense of the nature of the task. You can also use your results to construct as a key for comparison. Keep in mind, however, that the investigation requires interpretation and that your interpretation would be expected to differ from that of your students. A single "right answer" is not the goal of the investigation, so refrain from using your result as a "grading key" except for general properties of quality work like consistency with the evidence.

Photocopy **Blackline Master** *Our Dynamic Planet* 6.1, World Map of Continents and Continental Shelves for each student. Make extra copies available in the event students make errors while cutting up the continents or wish to revise their work. Have materials ready in advance. You can put glue (or glue sticks), scissors, construction paper, and photocopies in small trays or shoeboxes for easy distribution to student groups and collection at the end of the investigation.

Consider how you will review the content provided in the **Digging Deeper** reading section with students after they have done the investigation. Overhead transparencies from the following **Blackline Masters** might be helpful during class discussion:

Blackline Master *Our Dynamic Planet* 6.2, Ancient Mountain Belts [need to revise slightly – the mountain belt in Scandinavia has no number.]
Blackline Master *Our Dynamic Planet* 6.3, Generalized Distribution of Fossils of *Lystrosaurus*, *Glossopteris*, *Cynognathus*, and *Mesosaurus*
Blackline Master *Our Dynamic Planet* 6.4, Ice Sheet Distribution, 300 Million Years Ago
Blackline Master *Our Dynamic Planet* 6.5, Pangea
Blackline Master *Our Dynamic Planet* 6.6, Breakup of Pangea

A wall map of the world (optional) provides a nice visual aid that students can refer to easily during the investigation.

This might be a good point in the module to assess group participation, if you have not already done so. Suggestions for using the **Group Participation Evaluation Forms I and II** (available at the back of this Teacher's Edition) are provided within the investigation. Be sure to give each student a copy of the rubric and that you review the criteria before students start their work.

Materials
- a copy of the world map cutout showing the continents and the continental shelf
- scissors
- construction paper
- glue

Optional: Colored pencils (four different colors, or one color for each type of fossil evidence) for **Step 5** of the investigation. Alternatively, students can use a pencil and make a unique pattern for each type of evidence.

Investigation 6:

Earth's Moving Continents

Key Question

Before you begin, first think about this key question.

Have the continents and oceans always been in the positions they are today?

In **Investigation 4** you learned that the Earth's lithospheric plates move relative to one another. Do they go anywhere? How far do they move? Have they always been moving? Have there always been the same number of plates?

Share your thinking with others in your class. Keep a record of the discussion in your journal.

Materials Needed

For this investigation your group will need:

• a copy of the world map cutout showing the continents and the continental shelf

• scissors

• construction paper

• glue

Investigate

1. Look at the map of the world on the following page, centered on the Atlantic Ocean. Look especially at the edges of the African and South American continents.

Key Question

Begin by asking students to respond to the **Key Question**, "Have the continents and oceans always been in the positions they are today?" Tell students to not only record their response (yes or no) but also write down the basis of their ideas in their journal. After a few minutes, discuss their ideas in a brief conversation. Emphasize thinking and sharing of ideas. Avoid seeking closure (i.e., the "right" answer). Record all of the ideas that students share on an overhead transparency or on the chalkboard. Have students also record this information in their journals.

Student Conceptions about the Movement of the Continents

The permanence of continents and oceans is a deeply rooted conception. It took many years for geologists to accept the idea that continents move, or that ocean basins could open, close, and open again. Thus, you should anticipate some difficulties changing students' conceptions about the idea that continents and oceans move over time. At first, students may think of the question in terms of the shoreline (where the ocean and continents meet) or in terms of coastal erosion (yes, shorelines erode the land, and sea level rises and falls, which make coastlines migrate landward and seaward). They may also confuse the word continent with country. By the end of this investigation, students will be better able to think about this question in terms of plate tectonics.

Answer for the Teacher Only

Before the plate tectonics revolution, in the 1960s, only a minority of geoscientists believed that the continents have shifted their positions through geologic time. Now geoscientists recognize that the continents and oceans have shifted radically in their positions through geologic time. Before plate tectonics, it was very difficult to figure out how the geology of the continents had developed. Nowadays, ocean–continent subduction and continent–continent collision allow very natural and satisfying explanations for the geology of the continents. Just in recent years, with the advent of GPS (the satellite-based global positioning system), geoscientists can now measure directly the movement of North America away from Eurasia by several centimeters per year!

Investigation 6

About the Photo

The photograph of the headland at Point Lobos, California, provides a concrete example to help students consider the question about whether or not oceans and continents have always been in the position they are today. Point Lobos also provides an interesting example from this perspective of plate tectonics. Some 70 million years ago, the rocks found at the base of the sequence of rocks at Point Lobos (granodiorites—igneous rocks formed from magma that cools and crystallizes within the Earth) were located about 1600 km to the south near the tip of what is now know as Baja California, Mexico! Plate tectonic forces tore away a large block of the Earth's crust containing the granodiorite and moved it northward during several stages of tectonic activity.

Investigate

Teaching Suggestions and Sample Answers

Introducing the Investigation

In this investigation, students are asked to evaluate the geographic fit of continents on either side of the Atlantic Ocean and then cut out the outlines of continents and attempt to assemble them in a reasonable way into a single large continent. They are then asked to examine various lines of evidence in the geologic record of the continents (faunas and floras, mountain belts, and glacial features) to aid in the reconstruction. Tell students that each step of the investigation involves using some of the evidence that scientists gathered and used over a long period of time to develop an answer to the **Key Question** posed at the start of the investigation.

NOTES

INVESTIGATING OUR DYNAMIC PLANET

The dashed lines show the continental shelf, a shallow platform along the edge of all the continents.

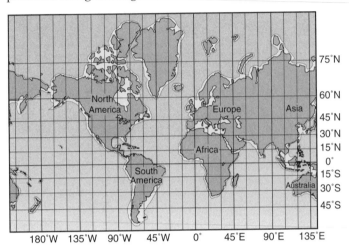

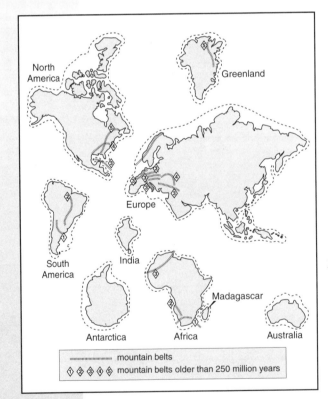

mountain belts

◊ ◊ ◊ ◊ ◊ mountain belts older than 250 million years

a) Describe the match between the East Coast of South America and the West Coast of Africa.

b) Describe the match between the bulge of West Africa and the outline of the East Coast of North America.

2. Examine the figure to the left that has all of the continents separated from one another so that they can be cut out with a pair of scissors.

Evidence for Ideas

1. a) The match is generally very good, although not perfect. The east coast of South America fits rather nicely into the west coast of Africa.

   b) The match is fairly good, but Florida and the Caribbean interfere with it. The reason is that much of the geologic development of that area postdates the breakup of Pangea.

2. This map shows the locations of mountain belts (depicted by an orange line) older than 250 million years. Five such belts are shown, each of which is numbered.

Conduct
Investigations

Get a copy of this map. Use scissors to cut out the continents along the outer edges of the continental shelves, along the dashed lines.

3. Use the cutouts of the continents like pieces of a jigsaw puzzle.

   On a sheet of construction paper, try to arrange the continents as one large landmass.

Evidence
for Ideas

   a) Describe the locations of any overlapping areas.

   b) How confident are you that the continents were linked together at some time in the past?

4. Several of the world's mountain ranges that appear on a continent today are similar in age and form to mountain ranges that today are on another continent. Some of these mountain ranges are shown in the continent cutouts. They are numbered according to those that have similarities with one another.

Evidence
for Ideas

   a) Do the mountain ranges with common features line up with one another in your arrangement of the continents?

3. As you circulate, check students' work to make sure that students are cutting the map up along the dashed lines. The dashed lines represent the edge of the continental shelf, which is typically at a depth of about 200 m.

    a) Answers will vary depending upon how the continents were arranged. However they are arranged, there will inevitably be some gaps and overlaps. Some students will probably be able to develop a better match than others. Without guidance provided by geologic features on the various continents (to come later in the investigation), this is a challenging exercise. There is still controversy among the experts about the exact geometry of Pangea. The only reliable aspect the students can use with some confidence is the match between Africa and South America across the South Atlantic.

    b) An honest student would have to admit that he or she could not be very confident of the reality of a single continental mass in the past.

4. Students may tend to simply match the "lines" that depict the belts. They need to pay attention to the numbers that differentiate each mountain belt (match a "number 5" on one continent with a "number 5" on another).

    a) Answers will vary: yes or no, depending upon how expert (lucky?) the student was in assembling the continents into one.

b) Does this information give supporting evidence for your arrangement of the continents, or does it argue against your arrangement of the continents?

Evidence for Ideas

c) Make changes in your model in light of the evidence you have.

5. Several fossils are found on particular landmasses but not on others. Review the following evidence:

Cynognathus was a reptile that lived in what are now Brazil and Africa.

Lystrosaurus was found in Central Africa, India, and Antarctica.

Megosaurus was found in the southern tip of South America and the southern tip of Africa

Glossopteris was a fern found in Antarctica, Australia, India, southern Africa, and southern South America

The map shows the locations of the fossils.

Inquiry

Using Evidence Collected by Others

In this investigation you are using evidence that you have been provided to formulate your ideas about how the continents may have fit together. Scientists must often rely on evidence gathered by others to develop their hypotheses.

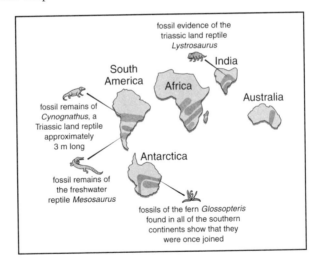

fossil evidence of the triassic land reptile *Lystrosaurus*

India

South America

Africa

Australia

fossil remains of *Cynognathus*, a Triassic land reptile approximately 3 m long

Antarctica

fossil remains of the freshwater reptile *Mesosaurus*

fossils of the fern *Glossopteris* found in all of the southern continents show that they were once joined

a) Add this information to the landmasses you are arranging on your construction paper.

b) Does this new information strengthen or weaken your model? Explain.

Consider Evidence

c) Make changes in your model in light of the new evidence you have.

d) What new information might change your model?

Seek Alternati

b) Answers will vary, again depending upon how close the assembly came to the generally accepted reconstruction of Pangea. Students now have two types of evidence to reconcile. In general, the addition of a new line of evidence (mountain belts older than 250 million years) will allow students to confirm or reject possible "fits" of continents that students had proposed earlier using only the fit at the edge of the continental shelf.

c) Some students will need to make more extensive changes than others. Emphasize to those whose assemblies were far off the mark that they had a very limited basis for doing the assembly in the first place, without the additional supporting evidence.

5. a) Students should carefully transfer the evidence from the diagram on page P54 to their maps. If you provide colored pencils and they use different colors to represent different kinds of fossil evidence, remind them to make a key on their map to show what each color represents.

b) This new information should end up strengthening all of the students' models.

c) These further changes are not likely to be as extensive as the changes made in **Step 4** (c).

d) An honest answer might be: any other geologic feature that allows a further geographic fit between continents. Some possibilities, about which most students will know little, like coals, evaporites, eolian (wind-blown) sand deposits, give evidence of past climatic belts. Students who read ahead to **Step 6** might note that glacial evidence might be useful.

6. Evidence produced by glaciers from long ago provides geoscientists with ideas about the movement of continents. Imagine a bulldozer plowing a pile of soil, then stopping, backing up, and driving away. Like a bulldozer, a glacier plows a large pile of rock and sediment. When the glacier melts, the "plowed" deposit, called a terminal moraine, is left at the front of the glacier.

Examine the next map, which shows where evidence of ice sheets 300 million years old has been found in the Southern Hemisphere. The red line on the map connects all the places where terminal moraines from this time have been found on the continents, and the arrows show the direction of glacier movement.

Inquiry

Sharing Findings

An important part of a scientific experiment is sharing the results with others. Scientists do this whenever they think that they have discovered scientifically interesting information. In this investigation you are sharing your ideas with other groups.

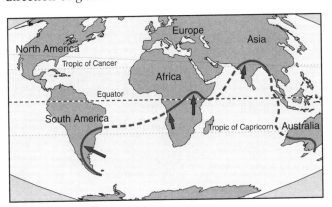

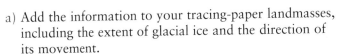

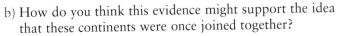

a) Add the information to your tracing-paper landmasses, including the extent of glacial ice and the direction of its movement.

b) How do you think this evidence might support the idea that these continents were once joined together?

c) Make changes in your model in light of the new evidence you have.

7. When you are satisfied with the model you have created, glue your landmasses onto the construction paper. Include a key or legend for all the information you have added.

a) What additional new information would you need in order to improve the model?

8. Share your model with the class. Discuss the evidence behind the model.

About the Photo

The red line on this diagram is described as the limit of end moraines. It might be more accurately described as the "limit of glacier ice." This limit was determined by looking at a variety of glacial evidence, including striated bedrock surfaces (places where rocks at the base of glaciers scratched the rocks over which the glacier moved) and a variety of glacial sedimentary deposits, not just end moraines.

---

6. The diagram may be misleading to students. Unless one assembles the continents into a single landmass (mentally, or physically, with paper cutouts as students are doing), it might create the impression that glaciers crossed the ocean from Antarctica. Remind students to interpret the diagram in the context of the fit of the continents.

   a) Students should transfer the red line (extent of glacier ice) and arrows (direction of movement of ice) onto their maps. They should add this information to their map key for future reference.

   b) Large continental ice sheets form on continent-scale landmasses located at high latitudes. The idea here is that, in order for an extensive ice sheet to have formed, all of the now widely separated continental areas must have been together as one large continental mass at a high latitude.

   c) There are likely to be further changes.

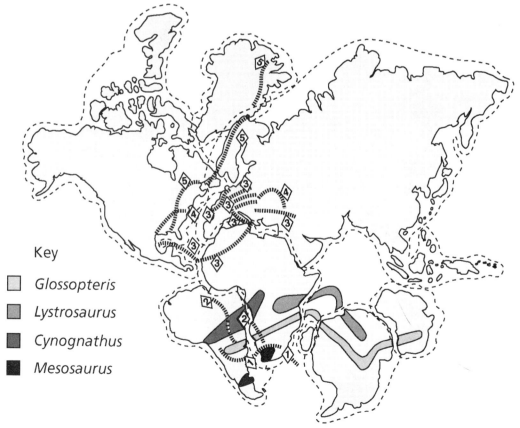

Key

☐ *Glossopteris*

▨ *Lystrosaurus*

▦ *Cynognathus*

■ *Mesosaurus*

Investigation 6

7. a) Answers will vary. A reasonable answer would simply be: more information of the same kind. A perceptive student might ask whether there is any information contained in the rocks of the continents that records the latitude and longitude of the continent in the past (geoscientists call that paleolatitude and paleolongitude). Certain kinds of rock do indeed carry such information, in the form of a record of the orientation and sense of the Earth's past magnetic field. That tells paleolatitude, although not paleolongitude. Such information has been of great value in deciphering the movements of continents, and pieces of continents, and in reconstructing past supercontinents. If any of your students would like to research this topic further, this field of geoscience is called paleomagnetism.

8. Allow students the opportunity to share and discuss their work. This is an important step in the investigation. It provides an opportunity for students to learn that different people may reach different conclusions from the same evidence. One method is to select several reconstructions (preferably ones that look different from one another) and make overhead transparencies from them. Have the groups that made those maps present their work at the overhead projector. Allow the class to question the group about their decisions for certain arrangements of the continents. If discrepancies or errors arise, be sure to allow students the opportunity to revise their work.

Assessment Tools

Group Participation Evaluation Sheets I and II
One of the challenges to assessing students who work in collaborative teams is assessing group participation. Students need to know that each group member must pull his or her weight. As a component of a complete assessment system, especially in a collaborative learning environment, it is often helpful to engage students in a self-assessment of their participation in a group. Knowing that their contributions to the group will be evaluated provides an additional motivational tool to keep students constructively engaged.

Group Participation Evaluation Sheets I and II provide students with an opportunity to assess group participation. In no case should the results of this evaluation be used as the sole source of assessment data. Rather, it is better to assign a weight to the results of this evaluation and factor it in with other sources of assessment data. If you have not done this before, you may be surprised to find how honestly students will critique their own work, often more intensely than you might do.

Journal Entry-Checklist
Use this checklist as a guide for quickly checking the quality and completeness of journal entries.

NOTES

Investigation 6

INVESTIGATING OUR DYNAMIC PLANET

Digging   Deeper

Evidence for Ideas

As You Read...
Think about:

1. *In your own words, explain the theory of continental drift.*
2. *What was Pangea?*
3. *How is a suture zone formed?*
4. *Why is the Pacific Ocean shrinking?*

SUPERCONTINENTS

Continental Drift

When you tried to assemble the continents like jigsaw-puzzle pieces, it probably seemed natural to you that Africa and South America fit together fairly well if you remove the ocean. This is one of the pieces of evidence that caused scientists 100 years ago to think that the two continents were once a single continent. The idea is that the single continent broke apart and the pieces drifted away from each other, to form the Atlantic Ocean. The fit of the continents is not the only evidence that supports the theory of continental drift. For example, you saw in your investigation that fossils of the same plants and animals are found in areas that are now separated by wide oceans and are in very different climatic zones.

Does it surprise you that it took a long time for most geoscientists to accept the theory of continental drift, even with the good evidence you worked with? The main reason was that no one could think of a way that the continents could plow along through the mantle beneath. When the theory of plate tectonics was developed in the 1960s, however, it gave a natural explanation for continental drift. Plate tectonic theory proposes that the outermost layer of the Earth, the lithosphere, behaves as a rigid layer. The lithosphere is broken into plates. These plates move relative to one another at their boundaries. Nowadays nearly all geoscientists believe in the reality of continental drift.

Supercontinents

In **Investigation 4** you learned that subduction can lead to the closing of an ocean and then continent–continent collision. When that happens, two separate continents

Digging Deeper

This section provides text and illustrations that give students greater insight into the evidence behind continental drift and the formation and breakup of supercontinents. You may wish to assign the **As You Read** questions as homework to help students focus on the major ideas in the text.

As You Read...

Think about:

1. According to plate tectonic theory, the outermost layer of the Earth consists of pieces of lithosphere called plates. The plates move relative to one another. Some of the plates have continents on them. That way, the continents move relative to one another.

2. Pangea was a continent that existed about 250 million years ago. It consisted of all of the Earth's continents, which had been gathered by continental drift into a single large continent, called a supercontinent.

3. A suture zone is the place where two separate continents come together to form a single continent. Suture zones form at continent–continent collision zones.

4. The Pacific Ocean is shrinking because there are subduction zones all around it. Oceanic lithosphere is being consumed at those subduction zones. There are mid-ocean ridges in the Pacific also, but the rate of production of new oceanic lithosphere there is less than the rate of consumption of oceanic lithosphere at the subduction zones, so the Pacific is shrinking.

Assessment Opportunity

You may wish to rephrase selected questions from the **As You Read** section into multiple choice or "true/false" format to use as a quiz. Use this quiz to assess student understanding and as a motivational tool to ensure that students complete the reading assignment and comprehend the main ideas.

become one large continent. Geoscientists are now sure that about 250 million years ago all of the Earth's continents were gathered into a single very large "supercontinent." That happened by a long series of continent–continent collisions. That supercontinent has been named Pangea (*pan* means "all", and *gea* means "land"). The diagram below is a map that shows geoscientists' best estimate of what Pangea looked like.

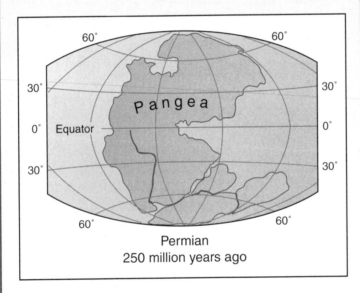

Permian
250 million years ago

You've already learned about some of the evidence for the existence of Pangea: for example, the fit of continents like Africa and South America, and similar fossils that are now far apart but must have been together in the past. Another important kind of evidence is the existence of former continent–continent collision zones, called suture zones, in the interiors of today's continents. These suture zones are the places where the earlier continents came together to form Pangea. The Appalachian Mountains, in eastern North America, are an example of these suture zones.

About the Illustration

This is a generalized map of what Pangea probably looked like. There are a number of versions of Pangea reconstruction in the geoscience literature, but they differ only in minor details. There is general agreement that Pangea looked about like this when it was fully assembled. Students may wonder why their reconstructions of past continents look so different from this one. The primary reason for the difference is that students used the current appearance of continents to construct their maps (which include portions of continents that did not exist 250 million years ago), whereas this diagram would have been constructed only upon the basis of rocks at least 250 million years old.

The Breakup of Pangea

About 200 million years ago the pattern of convection cells in the mantle changed, for reasons geoscientists are not yet sure about. This change caused Pangea to slowly split apart into several pieces. This process is called continental rifting. The pieces, which we know as today's continents, gradually drifted apart. That caused the Atlantic Ocean and the Indian Ocean, and the Antarctic Ocean to grow larger. The rifts didn't develop in exactly the same places where Pangea

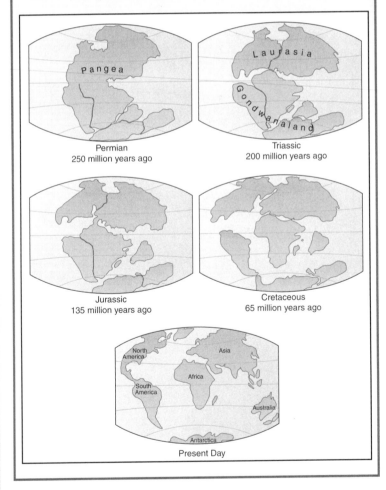

Permian
250 million years ago

Triassic
200 million years ago

Jurassic
135 million years ago

Cretaceous
65 million years ago

Present Day

About the Illustration

These sketch maps show a kind of "time-lapse movie" of how Pangea was broken apart into pieces and how the pieces drifted apart to result in the present-day configuration of the continents. Computer animations are available that show very graphically how this happened (links to these are available on the *Investigating Earth Systems* web site at www.agiweb.org/ies). The only exception to the general divergence of the continents since the breakup of Pangea is the collision of peninsular India with the southeastern margin of Asia. That collision, which has produced the Himalayas and the Tibetan Plateau, is still in progress.

was first sutured together. For example, the rift that formed the Atlantic Ocean was located to the east of the present Appalachian Mountains. That's why you sometimes hear that areas along the East Coast of the United States were once part of Africa! What does that really mean? They were on the *east* side of the ocean that vanished when northern Africa and North America were sutured together, but they were left on the *west* side of the new Atlantic Ocean that formed when Pangea was rifted apart.

There's evidence of earlier supercontinents, much farther back in geologic time. There seems to have been a supercontinent that formed and then rifted apart about 600 million years ago. Not nearly as much is known about the nature of that earlier supercontinent, because of the later movement of the continents while Pangea was being assembled.

At the time of Pangea, the Pacific Ocean was the world's only ocean! As the new oceans (the Atlantic, Indian, and Antarctic Oceans) have widened after Pangea was rifted apart, the Pacific Ocean has shrunk, although it's still the largest ocean. As you saw in **Investigation 4**, today's Pacific Ocean is surrounded by subduction zones. Those are the places where the floor of the Pacific Ocean is being consumed. What's going to happen in the geologic future? Will the Pacific continue to shrink, until all of today's continents collect there to form a new supercontinent? Or will the Pacific expand again, and the new oceans close up again to form a supercontinent where Pangea once existed? Most geoscientists think that the latter will happen.

Today scientists can actually measure how the plates are moving. They use orbiting satellites to directly measure the plates' movements as they happen. Only since the development of the satellite-based global positioning system (GPS) has this direct measurement of continental drift been possible.

NOTES

Investigation 6

INVESTIGATING OUR DYNAMIC PLANET

Review and Reflect

Review

1. Look again at the **Key Question** for this investigation: "Have the continents and oceans always been in the positions they are today?" In your journal, write down what you have learned from your investigations that provides an answer.

2. What kinds of evidence can be used to identify the former existence of a supercontinent?

3. In your own words, explain the theory of continental drift.

Reflect

4. Why do you think it took so long for most geoscientists to accept the theory of continental drift?

5. What additional evidence would you like to have to prove that the Earth's surface has moved and is moving?

6. What additional questions would you like to be answered and explained?

Thinking about the Earth System

7. What connections has your investigation revealed about the dynamic planet and the geosphere?

8. What links to the biosphere did you make in this investigation?

9. How was the hydrosphere connected to the evidence that you used?

   Don't forget to write any connections you uncover on the *Earth System Connection* sheet.

Thinking about Scientific Inquiry

10. How did you use evidence to develop scientific ideas?

11. How did you communicate your findings to others in a way that could be seen and understood?

Review and Reflect

Review

Your students should have gathered enough evidence to provide a reasonable answer to the **Key Question.** You need to emphasize that it is evidence that counts. They may have difficulty in understanding what constitutes evidence as opposed to opinion or inference. Take this opportunity to clarify both the nature and importance of evidence in scientific inquiry.

1. Students are likely to say that they have learned that there was a supercontinent called Pangea and that it was broken apart and the pieces drifted away to produce the present configuration of the continents. They might say that they have learned that the drift of continents is the result of the operation of plate tectonics, by which the continents are passive riders on lithospheric plates.

2. Students are likely to cite the kinds of evidence they used in reconstructing Pangea: geometrical fit of continents; matching of faunas and floras; matching of mountain belts; matching of evidence of continental glaciation.

3. According to plate tectonic theory, the outermost layer of the Earth consists of pieces of lithosphere called plates. The plates move relative to one another. Some of the plates have continents on them. That way, the continents move relative to one another.

Reflect

Allow students time to reflect on the nature of the evidence they have generated from their investigations. Again, help them see that evidence is crucial in scientific inquiry.

4. Students are most likely to note that although there was evidence that the continents were once joined, geologists had difficulty finding an acceptable method of moving the continents. There is a lot of inertia in science, and there is a lot of peer pressure among scientists as well. Most scientists work within the accepted paradigm, without spending much effort to "think outside the box." Despite what seems today as excellent evidence for the reality of continental drift, geologists were reluctant to admit or accept its reality, because nobody could envision a workable mechanism to explain it. It could even be dangerous to one's career to espouse continental drift! Eventually, geophysicists developed attractive theories for how the continents could move relative to one another, and it became more acceptable to believe in continental drift. Soon, almost everybody "jumped on the bandwagon." Within just a few years, there were only a few holdouts against the theory of continental drift.

5. A natural response would be that they would like to have direct measurements of the movements of continents relative to one another: if it's really happening,

maybe the movement can be measured. Many of your students are likely to know about GPS (the satellite-based global positioning system), because hand-held GPS receivers, which tell you exactly where you are on the Earth's surface, are becoming less and less expensive and more and more common. The precision of GPS has become so good in recent years that geoscientists can now measure the relative movement of continents, by several centimeters, from year to year.

6. Answers will vary. Encourage students to record at least one question, and remind students that the purpose of science is to raise questions as much as it is to answer them! The question should be one that could be researched as a homework or extra credit assignment.

Thinking about the Earth System

The knowledge students will have gained from this investigation presents a good opportunity for looking at the bigger picture of the Earth System. Help your students make as many connections as they can between their understanding of plate tectonics and the components of Earth's System. Remind them to look at the Earth System diagram on page Pviii at the front of their text.

7. Forces within the Earth, associated with plate tectonics, shape the configuration of the continents and oceans. Continent–continent collisions, which lead to assembly of supercontinents like Pangea, play a major role in shaping the geological record of the continents, because of the multitude of geological processes (earthquakes, volcanoes, rock deformation, mountain building) associated with the collision.

8. Plants and animals are adapted to their climatic environment. As a continent drifts into different climatic zones, the plants and animals must migrate in response to the drift or become extinct. The fossil record of where particular species of plants and animals once lived has been useful in figuring out how the continents have moved relative to one another.

9. Continental ice sheets develop only where there are extensive landmasses at high latitudes in climates where winter snowfall is greater than summer melting over large areas for long periods of time. The existence of glacial features on continents that today are in climatic zones not conducive to glaciation is evidence that the continents have drifted through geologic time. Also, students should have realized that the shapes of ocean basins have changed over time. They learned that the Atlantic Ocean did not exist 250 million years ago.

Teaching Tip

Remind students to enter any new connections that they have found on the *Earth System Connection* sheet. This might be a good time for students to review the entries they have made, as they are approaching the end of the module. Encourage students to reflect on the investigations they have completed and to connect what they have discovered to the Earth System. Are all the connections they have made entered on the sheet? Which connections are they having difficulty making?

Thinking about Scientific Inquiry

Help students to think about the inquiry processes they used in **Investigation 6**.

10. Evidence based on the geometrical fit of the continents and the matching of fossils, mountain belts, and glacial features between now-separated continents was used to develop models for the existence and configuration of the former supercontinent of Pangea.

11. Communication was accomplished by the use of maps showing the hypothesized configuration of Pangea.

Assessment Tool

Review and Reflect Journal Entry-Evaluation Sheet

Use the general criteria on this evaluation sheet for assessing content and thoroughness of student work. Adapt and modify the sheet to meet your needs. Consider involving students in selecting and modifying the criteria for evaluating their reflections on **Investigation 6**.

Investigation 6

Teacher Review

Use this section to reflect on and review the investigation. Keep in mind that your notes here are likely to be especially helpful when you teach this investigation again. Questions listed here are examples only.

Student Achievement

What evidence do you have that all students have met the science content objectives?

_____

Are there any students who need more help in reaching these objectives? If so, how can you provide this?_____

_____

What evidence do you have that all students have demonstrated their understanding of the inquiry processes?_____

_____

Which of these inquiry objectives do your students need to improve upon in future investigations? _____

_____

What evidence do the journal entries contain about what your students learned from this investigation? _____

_____

Planning

How well did this investigation fit into your class time?_____

_____

What changes can you make to improve your planning next time? _____

_____

Guiding and Facilitating Learning

How well did you focus and support inquiry while interacting with students?

_____

What changes can you make to improve classroom management for the next investigation or the next time you teach this investigation? _____

_____

How successful were you in encouraging all students to participate fully in science learning?_____

_____

How did you encourage and model the skills values, and attitudes of scientific inquiry? _____

_____

How did you nurture collaboration among students?_____

_____

Materials and Resources

What challenges did you encounter obtaining or using materials and/or resources needed for the activity? _____

_____

What changes can you make to better obtain and better manage materials and resources next time? _____

_____

Student Evaluation

Describe how you evaluated student progress. What worked well? What needs to be improved? _____

_____

How will you adapt your evaluation methods for next time?_____

_____

Describe how you guided students in self-assessment. _____

_____

Self Evaluation

How would you rate your teaching of this investigation? _____

_____

What advice would you give to a colleague who is planning to teach this investigation? _____

_____

Investigation 6

NOTES

INVESTIGATION 7: NATURAL HAZARDS AND OUR DYNAMIC PLANET

Background Information

Earthquake Hazards

Destructive earthquakes occur frequently in the United States and around the world. The hazards or effects of earthquakes can be divided into those directly related to the fault movement and passing waves and those hazards or effects that are triggered by the passing waves. Hazards caused directly by the passing waves and fault movement include ground shaking, fault rupture, ground subsidence, and tsunamis. Hazards triggered by the passing waves include fire, soil liquefaction, and earth movements like landslides, mudflows, and rock falls. Each of these hazards and the damage they cause is described below.

Ground shaking is caused by seismic waves as they travel along the Earth's surface. When the waves reach the surface they generate two kinds of surface waves. Damage from ground shaking is the most widespread and pervasive of all earthquake damage. Ground shaking can destroy buildings, bridges, and other structures and disrupt water, sewer, and utility lines. Additional damage from ground shaking results from objects falling from shelves and tables.

Fault rupture is caused by the spread of the fault plane to the surface. Depending on the orientation and sense of movement on the fault, the rupture may be evident as ground cracks or fault scarps. A fault scarp is a steep slope or cliff formed directly by movement along a fault before it is affected by erosion or weathering. Many of the steeper topographic features in the western United States, like the Grand Teton Mountains, are fault scarps. Numerous recent geologic events have caused ground cracking and fault rupture. For example, the 1906 San Francisco earthquake caused features to be offset by as much as three meters.

Ground subsidence is a lowering of the ground surface due to shifting of fault blocks, gravitational slumping, or landsliding. In Hawaii, the coastline was submerged 3.5 m during a magnitude 7.1 earthquake in 1975.

A tsunami is a sea wave produced by a large disturbance of the ocean floor. A tsunami can be triggered by large earthquakes, or by large landslides or volcanic eruptions. In 1946, an earthquake off the coast of Alaska triggered a tsunami that later arrived in Hawaii and produced waves 16 m high. The 1964 Alaska earthquake generated the most recent tsunami on the Pacific coast of the United States. That tsunami killed seven people in Crescent City, California. For large earthquakes that occur in the Pacific Ocean basin the earthquakes are located and tsunami warnings are issued to distant areas. Tsunamis move at about the speed of a jet airplane and can take up to 24 h to travel from one corner of the Pacific to the other. Unfortunately, some earthquakes are located close to land, and if they trigger a tsunami it arrives onshore very quickly with no time for a warning. In 1896, a locally generated tsunami in Japan killed 26,000 people.

Fires are often caused by the disruption of natural gas pipelines. Water lines are also broken, which makes fighting the fires difficult. Following the 1906 San Francisco earthquake, fires burned for three days and caused ten times more damage than the earthquake itself. New building codes have reduced the likelihood of large fires during California earthquakes.

Soil liquefaction is the transformation from a solid to a liquid state in a cohesionless soil as a result of increased fluid pressure in the pores in the soil. Soils and other soft geologic materials, like sand, with very high water content are also susceptible to seismically induced landsliding. Liquefaction has caused spectacular damage. During the 1964 earthquake in Niigata, Japan, large apartment buildings remained intact but toppled on their sides. Liquefaction of clay-rich soils during the 1964 earthquake in Alaska cut 300 m inland and 2800 m along the coast.

Landslides are downslope mass movements of soil, sediment, and/or rock material under the pull of gravity. Landslides vary widely in size, up to many cubic kilometers of material, and in speed, from imperceptibly slow to express-train speed. Landslides were numerous during the 1994 Loma Prieta earthquake in California. Landslides during the 1998 earthquake in Afghanistan destroyed entire villages, killing hundreds of people. Landslides also disrupt transportation routes, making disaster relief more difficult.

Volcanoes and Their Hazards

Differences in chemical composition cause the varying eruptive behavior of lavas. There is a broad spectrum of magma compositions, ranging from the silica-poor lava that forms most basalt, the most common rock on Earth's surface (although not necessarily the most common on the continents), to the silica-rich magma that forms rhyolite. This latter magma type tends to be highly explosive when it is erupted, because of the relatively high content of volatiles (dissolved substances with low boiling points, like water).

All magmas contain dissolved gases that are released into the atmosphere during and between eruptions. Water vapor is the most abundant gas released, followed by carbon dioxide and sulfur-containing gases (predominantly sulfur dioxide and hydrogen sulfide). Magma also releases minor amounts of other gases like hydrogen, carbon monoxide, hydrogen chloride, and hydrogen fluoride. The two volcanic gases most hazardous to people, animals, crops, and property are sulfur dioxide and carbon dioxide. Sulfur dioxide gas combines with water and oxygen to form sulfuric acid, and leads to abnormally acidic rain downwind from a volcano. Large explosive eruptions that inject enormous volumes of ash and sulfur dioxide gas into the stratosphere can contribute to temporary global cooling. Because carbon dioxide gas is heavier than air and collects in topographically low areas, people and animals in these low areas may suffocate and vegetation may die.

Explosive eruptions typically occur only along volcanic-island arcs and continental margins because magma generated in subduction zones is higher in silica than the basaltic magmas generated beneath mid-ocean ridges and at hot spots. The explosive eruptions of silica-rich ash can affect all of Earth's systems because the finest part of the ash circulates in Earth's atmosphere. A volcanic eruption can send a column of material many miles into the atmosphere. The 1991 eruption of Mt. Pinatubo in the Philippines was ten times more explosive than the 1980 eruption of Mt. St. Helens. It spewed out ten times the volume of volcanic products. Two million years ago an eruption at Yellowstone, Wyoming erupted more than a thousand times the volume of material released at Mt. St. Helens.

The mechanics of explosive volcanic eruptions are still not fully understood. When a body of magma moves upward in the Earth's crust, the confining pressure due to the weight of overlying rock decreases. If the magma contains no volatiles or only a low concentration of volatiles, as with basaltic magma, this decrease in pressure is not of great consequence, but if the magma contains a relatively high concentration of dissolved volatiles, like water and carbon dioxide, as is common with silica-rich magmas, then the volatiles have a tendency to come out of solution as the pressure is decreased. In some cases this release is gradual, but in other cases it is sudden. In such cases, when the magma is poised just below the surface a crack might propagate randomly through the overlying rocks down to the vicinity of the magma, allowing sudden release of pressure in the magma and very rapid outgassing of volatiles, causing a large and sometimes truly gigantic explosion. Geoscientists can make tentative predictions of the potential danger of explosions by monitoring such things as earth movements caused by the moving magma and gradual release of gases through fissures, but it is still not possible to make reliable predictions of the occurrence and timing of volcanic explosions.

The particulate material ejected from a volcano during an explosive eruption varies widely in composition. The most important kinds of particles are as follows: fragments of preexisting rock that lay above the magma before the eruption; crystals of various minerals that had already crystallized in the magma before it was erupted; rounded masses of very fine or glassy igneous rock that formed by chilling of drops or gobs of magma in flight; and curved shards of broken walls of gas bubbles. The size of the particles also varies widely.

Volcanologists use the term tephra for all of the ejected material, regardless of size or composition. Technically, the term ash refers only to the finest part of the tephra. Material of sand and pebble size is called lapilli, and the largest pieces are called blocks or bombs, depending on their shape. Blocks and bombs rain down close to the site of the eruption. (Imagine blocks the size of television sets falling out of the sky.) Great clouds of finer ash are put much higher into the atmosphere. In some photographs of explosive eruptions, you can see blocks and bombs falling out of the rising cloud of ash. Dispersal of the ash depends on the pattern of winds (speed and direction) at all levels in the atmosphere. Students should understand that the fall velocity, or settling velocity, of a solid particle in a fluid increases with the size and mass of the particle, so the finest part of the ash, of sizes down to the micron (0.001 mm) range, not only reach the highest levels of the atmosphere but travel the farthest with the wind before falling out. The finest part of the ash can stay in the atmosphere for many months, and during that time it is diffused throughout the entire atmosphere. Somewhat coarser ash falls out in a shorter time over areas that are subcontinental in size. Both the thickness and the particle size of the mantle of ash deposited on the land surface decrease gradually downwind of the eruption.

As with any liquid, lavas flow downslope in response to the pull of gravity. A given volume of lava in a flow is acted upon by the

Investigation 7

downslope component of the force of gravity. Newton's second law of motion would tell you that the fluid should be accelerated in its downslope motion. The reason it instead flows at an almost constant speed is that a friction force develops at the bottom of the flow, which counterbalances the downslope driving force of gravity. Lavas flow faster on steeper slopes because the downslope force of gravity is greater. Lavas that flow in channels tend to move faster than lavas that flow as wide sheets, because the area of the base of the flow, where the retarding force of friction is generated, is less in relation to the volume of the flow, to which the force of gravity is proportional. For the same reason, thicker flows of lava tend to move faster than thinner flows.

Pyroclastic flows are a very different kind of flow that is sometimes caused by an explosive eruption. If the volcano emits a large volume of a concentrated mixture of gases, liquid droplets, and solid particles, the mixture can flow downslope like a dense liquid. The effective viscosity of such a mixture is much less than the viscosity of a lava, so the speed of movement is much greater. Speeds of pyroclastic flows can be over 100 m per second, which is even faster than racing cars! The potential for destruction of plant and animal life, and human habitation, is staggering. When pyroclastic flows finally stop, they sometimes become welded into solid rock as the still-hot material cools.

Debris flows are similar to landslides, but they are different in that they consist of a high-concentration mixture of sediment and water that flows, as a fluid, downslope at low to moderate speeds. Large debris flows can fill valleys and bury villages. They are especially common in areas with steep slopes, deeply weathered rock, and occasionally heavy rainfall. If the concentration of solid particles in a debris flow is sufficiently high,

greater than 40% to 50% by volume, the mixture of water and particles tends to flow like a viscous liquid. Because of the high concentration of solid particles, they cannot readily settle out to the bottom of the flow, as would be the case in an ordinary river flow, so the mixture can flow for long distances. If any of your students have had experience with concrete, they could appreciate that debris flows have a consistency something like that of fresh concrete that has just a bit too much water. Speeds of debris flows vary from very slow, no faster than a walk, to very fast, tens of meters per second. Areas susceptible to debris flows are those where slopes are steep and loose sediment containing fine as well as coarse materials is common. Land surfaces with steep slopes and a mantle of weathered volcanic ash are particularly susceptible to debris flows.

Debris flows whose solid materials are mainly or entirely volcanic ash are called lahars. Lahars commonly develop during heavy rains some time (often a long time) after an explosive eruption has covered the land with a thick layer of volcanic ash. Lahars are especially common after weathering of the ash in a warm, humid climate has produced abundant fine clay material in the layer. Debris flows can start in various ways: runoff down slopes during and after especially heavy rainfall; collapse and sliding of steep, water-saturated slopes; or breakout of temporary lakes in terrain covered by volcanic ash. Debris flows, including lahars, tend to find their way into preexisting stream or river valleys, and when they finally come to a stop, the valley is filled with a deep layer of watery sediment. Entire villages can be buried almost instantly in this way. The village of Herculaneum, near the modern city of Naples in Italy, was deeply buried by a lahar that resulted from an eruption of Vesuvius two millennia ago.

More Information...on the Web
Go to the *Investigating Earth Systems* web
site www.agiweb.org/ies for links to a variety
of other web sites that will help you deepen
your understanding of content and prepare
you to teach this module.

Investigation Overview

The connection between the geosphere and biosphere is the primary focus of this culminating investigation. Students use their knowledge and skills about *Our Dynamic Planet* to devise an information brochure for a community that is located near an earthquake-hazard or volcano-hazard site.

Goals and Objectives

The goal of this investigation is to deepen students' knowledge and understanding of the natural hazards that stem from our dynamic geosphere, and to appreciate the value of Earth science understanding and information to minimizing human and property loss. The investigation encourages students to review science content and inquiry processes that have been used throughout the module. It can be used as a final assessment, review for a final test, or both. As a result of **Investigation 7**, students will develop a better understanding of the natural hazards associated with earthquakes and volcanoes, and how communities can prepare for these events. They will also improve their ability to communicate scientific information effectively to others.

Science Content Objectives

Students will collect evidence that:
1. Natural hazards have an impact on communities.
2. Natural hazards like earthquakes and volcanoes are more likely to happen in some areas than in others.
3. If well informed, humans can prepare themselves for natural disasters.

Inquiry Process Skills

Students will:
1. Use data and information about natural disasters to create a brochure.
2. Analyze which data are suitable for inclusion in the brochure.
3. Conduct additional research on natural disasters.
4. Collate information into a useful format.
5. Communicate to others what they have discovered about natural disasters and how to deal with them.

Connections to Standards and Benchmarks

In this investigation, students explore natural hazards associated with earthquakes and volcanoes. All the content standards and benchmarks that students have been working toward understanding come together in this final investigation. Remember, these are statements of what students are expected to understand by the time they complete eighth grade. What they have been doing throughout this module on *Our Dynamic Planet* is just part of that ultimate learning outcome. Your students will have developed their understanding of some of these ideas, at least in part. However, many students will require additional experiences.

As your students work through **Investigation 7**, keep these standards and benchmarks in mind and note the general level of understanding evident in what students discuss and do. Do not attempt to "teach" these standards directly. The role here is to guide students from the ideas they have toward a more complete understanding.

NSES Links

• Natural hazards include earthquakes, landslides, wildfires, volcanic eruptions, floods, storms, and even possible impacts of asteroids.

• Students should understand the risks associated with natural hazards (fires, floods, tornadoes, hurricanes, earthquakes, and volcanic eruptions), with chemical hazards (pollutants in air, water, soil and food).

AAAS Links

• The interior of the Earth is hot. Heat flow and movement of material within the Earth cause earthquakes and volcanic eruptions and create mountains and ocean basins. Gas and dust from large volcanoes can change the atmosphere.

• Some changes in the Earth's surface are abrupt (such as earthquakes and volcanic eruptions) while other changes happen very slowly (such as uplift and wearing down of mountains). The Earth's surface is shaped in part by the motion of water and wind over very long times, which act to level mountain ranges.

Preparation and Materials Needed

Preparation

Your students will need to come up with their own plan for making an information brochure about natural hazards. You will want to have resources—newspaper articles, books, Internet access, etc.—available for them to do some additional research. One of the most valuable resources for your students will be the notes they have collected in their journals. The exact materials that need to be gathered to make an information brochure about natural hazards depend upon what you and your students decide to do. Here are some ideas for presentations, beyond what is given in the student text:

- booklet about your area;
- brochure;
- poster (perhaps with detailed information on the back);
- flowchart;
- display;
- video program;
- guided "audio car" tour of the region;
- illustrated presentation that can be given to an audience;
- web site.

Decide on the composition of student groups. Because this is the last investigation of the module, you may want to use it as an assessment tool or as a review for a final test (or both). Choose groups carefully so that there is a mix of abilities and good group dynamics. If the group dynamics are not good, change them before the end of the first part of **Investigation 7**.

The table on page P62 lists 11 different natural hazards associated with earthquakes and volcanoes. One option is to assign hazards to student groups and for students to conduct generalized research about the hazards in various geographic settings. A second option is to focus upon particular places where earthquake and volcano hazards exist and to let the location guide students' research. Examples of such locations are provided at the *Investigating Earth Systems* web site.

Make overhead transparencies of the following **Blackline Masters**, found at the back of this guide for use during class discussion:

Blackline Master *Our Dynamic Planet* 7.1, Earthquake and Volcano Hazards
Blackline Master *Our Dynamic Planet* 7.2, Explosive Volcanic Eruption

Materials

- reference materials *
- access to computer word processing and desktop publishing (if possible)
- range of general craft materials required to make a brochure

* The *Investigating Earth Systems* web site www.agiweb.org/ies/ provides topical Internet sites and a list of resources that will aid student research.

NOTES

Investigation 7:

Natural Hazards and Our Dynamic Planet

Putting It All Together

Explore Questions

Key Question

Before you begin this final investigation, first think about this key question.

What natural hazards do dynamic events cause?

Think about the movement of Earth's lithospheric plates. What hazards can they pose to humans? Think about all you have learned in the previous investigations. Share your thinking with others in your class. Keep a record of the discussion in your journal.

Materials Needed

For this investigation, each group may need:

• reference materials

• access to computer word processing and desktop publishing (if possible)

• range of general craft materials required to make a brochure

Investigate

1. Consider what might happen if an earthquake or volcanic eruption occurred in or close to a community. In your group, brainstorm the ideas that you have about the questions below:

[handwritten: Include article of Commentary tables]

Key Question

[handwritten: alt - Do GIS Module 2 Chap 2 (Life on the Edge)]

Have students respond to the **Key Question**, "What natural hazards do dynamic events cause?" By now, your students should have a good understanding of all the aspects of *Our Dynamic Planet* that they need to apply here. It might be helpful to hold a brief discussion and summary about each point (see the list in **Investigate, Step 1**) emphasizing to students that the knowledge they have gained in all of the previous six investigations will be crucial in successfully completing **Investigation 7**.

Student Conceptions of Natural Hazards Caused by Dynamic Events

Middle-school students often note that earthquakes are natural hazards that cause damage to buildings, homes, street, and bridges, and that the collapse of structures may harm or kill people. Students most commonly cite lava as a natural hazard from volcanic eruptions. Students are less likely to know or realize other hazards are associated with earthquakes and volcanic eruptions, or that the hazards may persist for some time after an event (lahars can be a problems for months, sometimes years, after a volcanic eruption). For example, they may not know that earthquakes may break water mains and gas lines, or that if a fire results from the event, firefighters may not be able to fight the blaze. Few students realize that in addition to lava, volcanic ash and lahars present a variety of hazards during eruptions.

Answer for the Teacher Only

Refer to **Background Information** for this investigation, as well as the **Digging Deeper** reading section.

About the Photo

Advancing lava will burn everything in its path. In this case, eruption of fluid basaltic lava in Hawaii flowed into a village and ignited trees and homes. Lava flows downhill under the force of gravity. Note how the road slopes from left to right, and how the cooling lava (black mass in lower left side of the photograph) is flowing down the street ("downhill") to the right.

Assessment Tool

Key Question Evaluation Sheet
Use this evaluation sheet to help students understand and internalize basic expectations for the warm-up activity.

Investigate

Teaching Suggestions and Sample Answers

1. Asking students to think about what would happen if an earthquake or volcanic eruption occurred in their community is a way to make them think more deeply about the potential impact of these events on human life and property.

Assessment Tools

Assessing the Final Investigation

To complete **Investigation 7**, students need a working knowledge of the activities carried out in the previous six investigations in this module. Because it refers to the previous steps, **Investigation 7** is a good review and a chance to demonstrate proficiency. For an idea of how to use the last investigation as a performance-based exam, see the section in the back of this Teacher's Edition. If you chose to use a scoring guide, review it with students before they begin their work.

Student Presentation Evaluation Form

Use the **Student Presentation Evaluation Form** as a simple guideline for assessing presentations. Adapt and modify the evaluation form to suit your needs. Provide the form to your students and discuss the assessment criteria before they begin their work.

NOTES

Investigation 7

INVESTIGATING OUR DYNAMIC PLANET

Explore
Questions

a) What would happen to local dams, reservoirs, or water supply stations?

b) What would happen to schools, homes, and government buildings?

c) What would happen to electrical power plants?

d) What would happen to local hospitals?

2. Share your ideas in a class discussion.

Do not be too concerned at this point whether or not your initial ideas are correct. Just ensure that all ideas are given. The goal of this investigation is to learn about natural hazards. Pay careful attention to what your classmates contribute to this discussion.

3. Here are just some of the hazards that can be caused by volcanoes and earthquakes.

Event	Effect	Examples
Volcanoes	Eruption with lava flow	• lava streams burn all in their path
	Eruption with ash fall	• aircraft endangered, roofs collapse
	Lahar (mud flow)	• Large, fast-moving river of mud
	Pyroclastic flow	• hot mobile flow of volcanic material
	Lateral blast	• explosive wave knocks down all in its path
	Volcano collapses	• land drops away— homes threatened
	Volcanic gases	• volcanic pollution
Earthquakes	Ground motion	• buildings and bridges collapse
	Fault displacement	• roads crack, rail tracks split
	Fires	• ruptures in gas lines cause building fires
	Landslides	• shaking causes rock to slide downhill
	Liquefaction	• ground becomes like quicksand
	Failures of dams	• flooding

1. a) Failure of dams is common during a major earthquake. This results in loss of water supply, and it is likely to cause serious loss of life by flooding downstream of the dam. Water mains often break during an earthquake, making firefighting difficult and necessitating emergency supplies of water for cooking and drinking.

   b) Many buildings are damaged during a major earthquake, especially older buildings that were not built in accordance with more recently developed building codes. Damage ranges from minor cracks to complete collapse. Sometimes the soil beneath a building is liquefied by the shaking, and the building founders into the soil, often tipping over at a large angle. Large multistory buildings that are not built to code sometimes "pancake," with the floors collapsing one upon another like an accordion.

   c) Damage to power plants can result in widespread loss of power. Even if the generating plant is not damaged, power lines are often severed. The effect upon the community can be significant, including loss of heating and cooling capacity in homes. People in homes with private water wells lose their water supply during a power outage because groundwater is commonly pumped to the surface using electrical pumps.

   d) Aside from structural damage, the problem of sudden loss of electrical power is a serious problem. Most hospitals have emergency power generating capability, which can be started up and switched on immediately to maintain life support systems.

2. Have students share their ideas. Encourage students to speculate about the possibilities. If students' ideas are challenged by others, remind the class that the questions should be resolved through their research, not from you at this stage.

3. You may wish to take some time to discuss the hazards with students. Students who may have had firsthand experience with any of these hazards may wish to volunteer their insight. However, be

Teaching Tip

You may wish to display an overhead transparency of **Blackline Master** *Our Dynamic Planet* 7.1, Earthquake and Volcano Hazards for students to refer to as they review the nature of the assignment.

Investigation 7

Design Investigations

4. Using the resources available in your class, school media center, public library, and home, select one of the terms to research. Find out:

- what the hazard is;
- how it forms and works;
- how it affects living and non-living things;
- what steps citizens can take to prepare for such hazards;
- what people can do to protect themselves once the hazard starts;
- where people can get further information.

Here are some of the factors you could consider when assessing potential earthquake hazards for a particular area, or designing an investigation into earthquake hazards:

- closeness to active earthquake faults;
- seismic history of the region (how often earthquakes occur; time since last earthquake);
- building construction (type of building and foundation; architectural layout; materials used; quality of workmanship; extent to which earthquake resistance was considered by the designer);
- local conditions (type and condition of soil; slope of the land; fill material; geologic structure of the earth beneath; annual rainfall).

4. Note that the important aspects of earthquake hazards listed here (and aspects of volcano hazards listed on page P62) pose critical and challenging scientific questions. The goal of this activity is not for students to create a fancy brochure that warns people about hazards but for students to deepen their understanding of the science behind natural hazards and natural processes. It is important to hold students accountable for these deeper scientific questions.

Teaching Tip

The **Digging Deeper** reading section should be the first source of information for students to consult when researching a natural hazard. To ensure that all students have a basic introduction to earthquake hazards and volcano hazards, assign the reading and the **As You Read** questions for homework.

INVESTIGATING OUR DYNAMIC PLANET

Here are some of the factors you could consider when assessing potential volcanic hazards for a particular area:

- volcanic history of the area (how often eruptions occur; time since last eruption?);

- population of the area around the volcano (are there towns and villages in high risk areas?);

- prevailing wind directions (where will most of the volcanic ash settle?);

- topography of the land (where are there valleys and ridges that will direct where the lava and hot gasses flow out of the volcano?).

5. It is important that you find out all you can about your chosen hazard.

To do this, you may want to divide up the tasks, with each member of your group specializing in a particular aspect, using all information sources available.

a) List the responsibilities of each group member in your journal.

6. When you are sure you have organized your research in a reasonable way, begin your investigation.

When each person has conducted his or her research, share and discuss your findings in your group.

You may think that you need to experiment further, to establish clearly how your hazard works. If necessary, design and model your hazard, using readily available materials.

7. Once you have assembled all the information you need, and completed any tests you think necessary, design a brochure.

The job of the brochure is to provide information to residents of a community that is close to a potential earthquake or volcano hazard site. Be sure the brochure addresses all the points in **Step 4.**

Discuss the best way to organize your information to cover these points.

Here are a number of ideas to consider when designing your brochure:

- the shape and size of the brochure;

Inquiry

Dividing Tasks

This investigation provides you with an opportunity to mirror the teamwork that often happens in scientific studies. Different scientists often take on responsibility for different parts of a study.

Presenting Information

Scientists are often asked to provide information to the public. In doing so, they need to consider both the message they want to communicate and the persons or groups that will be using the information. Once they are clear about the message and the audience, they can then decide on the best method of presenting the information. These are decisions you also will need to make in this investigation.

Design
Investigation

Conduct
Investigation

Collect &
Review

Consider
Evidence

5. You and your class will need to decide whether or not to have groups specialize in different kinds of natural hazards. How you decide to do this will ultimately depend upon the size of your class, number of groups, and materials available for research.

   a) Recording roles and responsibilities increases the probability that students develop a clear, fair, and organized work plan at the outset of a project. It will also reduce the likelihood of students being dissatisfied about the amount of work they had to do compared with other members of their group. Point out how the work plan models the process that scientists engage in when working as a team (i.e., within a collaborative research project).

6. Suggest that students create an outline. Insist that all students keep a copy of the outline in their journals. Ensure that they refer to their journals throughout the investigation.

   It is important that students see the connection between the investigations they have done earlier (especially the information they have gathered in the process) and the application of these ideas in their information brochure.

Teaching Tip

If you encourage or require students to make a physical model of their natural hazard, review their plans and basic safety precautions before allowing them to proceed. If time and materials prevent you from having students do this during class, consider whether or not this presents an opportunity to engage your students' parents in the inquiry process by asking students to do the modeling at home under adult supervision.

7. The information provided in the Student Book focuses upon a paper brochure. Obviously other criteria and suggestions should be provided to students if they are asked to create a different type of product (web site, audio tour, etc.).

- the color of the paper or card that can be used;
- computer programs that have templates for brochure design;
- artwork, diagrams, charts, or drawings that you can use or create;
- the various talents members of your group have.

Keep in mind that the brochure has size limits and that you may need to find creative ways to include all the information you think is essential.

Work together to produce the best brochure you can, keeping in mind the audience for which it is intended.

8. When all group's brochures are complete, arrange a session where groups look at each brochure in turn.

 Digging **Deeper**

EARTHQUAKE HAZARDS

Earthquakes happen when there is sudden movement of two rock masses along a fracture plane called a fault. Because of large-scale movements of the Earth's lithospheric plates, great forces can build up in rocks. Eventually, when the forces become greater than the strength of the rocks, long fractures form, and the rocks on either side of the fracture surface shift relative to one another. This motion is jerky and irregular, which causes strong vibrations. The vibrations travel away from the fault in the form of seismic waves. Above the fault, the vibrations cause up-and-down motions and side-to-side motions of the ground surface. Those motions are what you feel as an earthquake. Earthquakes vary greatly in their strength. Most earthquakes are so small that they can be detected only with special instruments. Some earthquakes, however, release an enormous amount of energy. They can cause ground motions so strong that people who are out in the open can't even stand up!

As You Read...
Think about:
1. What causes earthquakes?
2. What effects can earthquakes have on buildings?
3. What is the hazard associated with liquefaction?
4. What is the difference between an ash fall and an ash flow?
5. What warning signs make volcanic eruptions easier to predict than earthquakes?

8. If you decide to assess students' presentations (see **Assessment Tool** below) be sure to review the criteria for such evaluation before students begin to prepare their presentation. You might assign a particular value or weight to the presentation, and the remainder to the specific product that they create. You can combine these components for the final unit assessment.

   After the presentations, it is important that you allot class time for students to discuss what impact their research has had on them personally. Help them to explore how the knowledge they now have has changed their views about natural hazards, and how they might change their behavior because of it.

Teaching Tip

Investigation 7 is purposely "open-ended" and not prescriptive. Your students have a number of choices they can make. The key assessment issue here is the quality of those choices, especially the level to which they are informed by the evidence and explanations they have derived from earlier investigations.

Assessment Tool

Student Presentation Evaluation Form
Use the **Student Presentation Evaluation Form** as a simple guideline for assessing presentations. Adapt and modify the evaluation form to suit your needs. Provide the form to your students and discuss the assessment criteria before they begin their work.

Assessment Strategy

Peer review is an important part of the scientific process. Help students provide constructive feedback about their colleagues' presentations by providing sentence stems that they should complete for their reviews. Examples include:
This presentation was effective at showing...
This presentation helped me to understand...
This presentation needed work on...
To improve this presentation, I would suggest that you...

Digging Deeper

This section provides text and an illustration that give students greater insight into natural hazards associated with earthquakes and volcanoes. You may wish to assign the **As You Read** questions as homework to help students focus on the major ideas in the text.

As You Read...

Think about:

1. Forces that build up in rocks sometimes become larger than the strength of the rock, and then the rock breaks along a fracture plane called a fault. The sliding of the rocks along the fault plane causes vibrations in the rock, which is felt as an earthquake.

2. Earthquakes can cause buildings to collapse. Sometimes the floors of a building collapse on one another. Sometimes the foundation of the building gives way, and the building tips over.

3. When the soil is liquefied, it can't support solid objects like buildings. The buildings can sink into the soil while it is liquefied.

4. An ashfall happens when a volcanic eruption sends particles of ash high into the atmosphere. The ash then falls out through the air to form a layer of ash over a large area of the ground surface. An ashflow happens when the volcano erupts such large quantities of ash that the column of ash collapses back downward, and the mixture of ash and gases flows down the slope of the volcano at a high speed.

5. Gases begin to escape from the ground around the volcano. Minor earthquakes in the vicinity of the volcano become more common. Sometimes, the volcano bulges upward slightly before an eruption.

Assessment Opportunity

You may wish to rephrase selected questions from the **As You Read** section into multiple choice or "true/false" format to use as a quiz. Use this quiz to assess student understanding and as a motivational tool to ensure that students complete the reading assignment and comprehend the main ideas.

NOTES

The most serious hazard associated with earthquakes is the collapse of buildings. If a large building is not carefully designed to withstand the shaking of an earthquake, the floors can collapse upon one another in a kind of "pancaking." If the foundation of a building is not adequate, the building can tip over sideways during a strong earthquake. Structural engineers continue to make careful studies on how to design buildings to withstand earthquakes. Cities in areas of the United States that are prone to earthquakes have building codes that builders are required to follow. Loss of life in earthquakes in the United States is much less than in some other countries where buildings are not designed to withstand large earthquakes.

Earthquakes can cause many other kinds of serious damage. In areas with steep land slopes, earthquakes can trigger large and fast-moving landslides. Water mains can be broken, making it difficult to fight fires, which are often caused by earthquakes. In areas where the soil is porous and saturated with water, the earthquake vibrations sometimes cause the material to settle into closer packing of the soil particles. When that happens, the soil can flow like a liquid. The process is called liquefaction. Liquefaction can cause buildings to sink into the ground!

About the Photo

The movement of the ground during an earthquake has caused the walls of a home to weaken and buckle, collapsing the second floor of the structure onto the garage at ground level.

Volcano Hazards

Lava flows from volcanoes are not especially hazardous to human life, because they flow slowly enough for people to get out of their way. Of course, they burn buildings that are in their path! Explosive volcanoes are far more dangerous. Such volcanoes throw enormous quantities of rock and mineral particles, called volcanic ash, high into the atmosphere. The ash settles back to the ground for distances of tens to hundreds of miles. The ash fall forms a blanket as thick as several meters on the ground surface. Even a thin blanket of ash can collapse roofs and kill crops. By far the worst hazard associated with explosive volcanoes, however, is an ash flow. Sometimes, volcanoes erupt ash in such large quantities that it collapses back downward over the volcano and rushes down the slopes of the volcano as a thick mixture of hot ash and volcanic gases. Ash flows move at express-train speeds of hundreds of meters per second. They can travel for tens or hundreds of miles, killing and burying everything in their path.

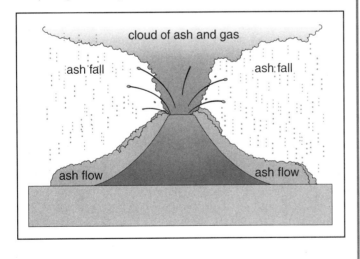

cloud of ash and gas

ash fall — ash fall

ash flow — ash flow

Teaching Tip

You may wish to use **Blackline Master** *Our Dynamic Planet* 7.2, Explosive Volcanic Eruption to make overheads to use when discussing the **Digging Deeper** reading section. One interesting aspect of this schematic diagram is that it assumes that the wind is not blowing. Ask students to describe what the cloud of ash as well as the pattern of ashfall would look like given a more realistic scenario, where wind blows in a particular direction for a period of time. Another aspect to the diagram is that the ashflow is distributed around the entire volcano (although in cross-section view, it looks like only two sides). Again, have students think about how the ashflow might be different if it comes down only one side of the volcano. Thus, the diagram raises a broader question – are all regions around a volcano equally subject to a hazard, or are some sides more likely to be at risk than others? For volcanic ashfall, scientists consider the prevailing winds and assign greater risk to communities downwind of volcanoes than those located upwind.

INVESTIGATING OUR DYNAMIC PLANET

Predicting Earthquakes and Volcanoes

Many attempts have been made to develop ways of predicting earthquakes. So far, no reliable method has been developed. Geoscientists are not sure whether they will ever be successful in predicting earthquakes with a high degree of certainty. If a weather forecast of a major snowstorm is wrong, and it rains instead, that's not a big problem. If there is a forecast of a major earthquake, large numbers of people might be evacuated from a city. If the earthquake didn't happen, think of all the unnecessary economic disruption there would be.

Volcanic eruptions are easier to predict, because most volcanoes give warnings of an eruption. Gases begin to escape from around the volcano. Minor earthquakes in the area beneath the volcano become much more common. Often, the surface of the volcano swells upward enough to be measured by surveying methods. At that point, a major eruption is likely, although not certain. Evacuations based on such observations have saved many lives.

About the Photo

The photo shows damage during the Kobe, Japan, earthquake. Ask students to identify the hazards and safety issues apparent in this photograph. (Students may identify: fire, downed electrical lines, unstable material that may fall off buildings in aftershocks, and blocked roads which prevent access for rescue teams, to name a few.)

Investigation 7: Natural Hazards and Our Dynamic Planet

Review and Reflect

Review

1. Describe an example of an earthquake hazard.

2. Describe an example of a volcanic hazard.

3. How do ash falls and ash flows from volcanic eruptions pose hazards to humans?

Reflect

4. What are the advantages and disadvantages of attempting to predict earthquakes and volcanoes?

5. The movement along a fault that causes an earthquake can also damage gas lines and water lines. Why does this pose a problem after an earthquake?

6. Why do you think people live in regions that are prone to volcanic eruptions and earthquakes?

7. Why is the kind of information that you provided for your report useful to the public?

Thinking about the Earth System

8. How might a volcanic eruption affect the biosphere, hydrosphere, or atmosphere in a negative way?

9. How do volcanic eruptions benefit the biosphere, hydrosphere, or atmosphere?

Thinking about Scientific Inquiry

10. Why do scientists present their findings to others?

11. What are some of the advantages to doing scientific work as a team?

12. What inquiry processes did you use in this final investigation? Name at least three processes, where you used them, and how they helped you complete this assignment.

Review and Reflect

Review

From their modeling and their research, your students should have gathered enough evidence to provide a reasonable answer to the **Key Question**. Again, emphasize that it is evidence that counts. From their investigations, your students should be able to understand important natural hazards that dynamic events cause.

1. Answers will vary. Several earthquake hazards are mentioned in the **Digging Deeper** reading section: damage to buildings, and even collapse; landslides; breakage of water mains; liquefaction of the soil beneath man-made structures.

2. Answers will vary. The two most serious volcano hazards are mentioned in the **Digging Deeper** reading section: ashfalls and ashflows. Students will probably already know that a lava flow burns everything in its path. It is less likely that they will be aware that many volcanoes emit toxic gases, which can asphyxiate or poison people and animals.

3. Ashfalls, if they are large enough, can suffocate humans and animals. Ashfalls can cause collapse of roofs of buildings, and they bury crops. Ashflows are an even greater hazard. They move at express-train speeds, and they can bury large areas of the land surface almost instantly.

Reflect

Give students time to reflect on the nature of the evidence they have generated from their investigations. Again, help them see that evidence is crucial in scientific inquiry. Use this session to pull together all that students have learned.

4. The obvious advantage is that people have warning and can take protective measures or be evacuated. The problem is that methods of prediction are still very uncertain. If a prediction is not made and the event happens, the organization making the prediction might be held liable for negligence. If a prediction is made but the event does not happen, the organization making the prediction is blamed for unnecessary disruption. If a prediction is made far in advance, commercial interests in the area might claim economic injury resulting from loss of business.

5. Gas lines can catch fire. Water lines are needed for fighting fires, which are common immediately after an earthquake. In large population enters, loss of water supply means that water must be trucked in on an emergency basis for cooking and drinking.

6. Human nature has a lot to do with it. There is a strong tendency for people to go on with their daily lives and not think far ahead about the unpleasantness of the distant possibility of an earthquake or volcano sometime in the future.

Investigation 7

7. Planners need to be aware of what might happen during a natural disaster, so that they can plan against it better. The general public needs to be aware as well, so that they can plan ahead if they choose to do so.

Thinking about the Earth System

This investigation, and the knowledge that students will have gained from it, presents a good opportunity for looking at the bigger picture of the Earth System. Help your students make as many connections as they can between their understanding of natural hazards and the Earth System.

8. A major volcanic eruption, especially an explosive one, kills plants and animals directly over a wide area. Temporary climate change worldwide might also cause environmental stress on certain kinds of plants and animals over continent-scale areas. Fine volcanic ash put high into the atmosphere during a major explosive volcanic eruption tends to block sunlight for months or even years, causing temporary global cooling. Effects on the hydrosphere are less direct, having to do with temporarily changed patterns of precipitation.

9. Volcanic ash tends to weather rapidly into rich soils for agriculture.

Teaching Tip

Remind students to enter any new connections that they have found on the *Earth System Connection* sheet. Encourage them to check to make sure all the connections they have discovered so far are entered on the sheet.

Thinking about Scientific Inquiry

10. Science is supposed to be about advancing human understanding of the world. All scientists need to know the research that has already been done, in order to guide their thinking and to plan new research. This is done in various ways: publishing books and articles; making presentations at meeting and conferences; and just talking informally with colleagues, in person, over the telephone or by e-mail.

11. There's an old saying (and it's usually true) that two heads are better than one. New ideas and approaches in science usually develop more readily when two or more scientists engage in discussions. It's not nearly as easy to see a shortcoming or flaw in your own thinking as it is to see that in someone else's thinking!

12. Answers will vary.

Assessment Tool

Review and Reflect Journal Entry-Evaluation Sheet

Use the general criteria on this evaluation sheet for assessing content and thoroughness of student work. Adapt and modify the sheet to meet your needs. Consider involving students in selecting and modifying the criteria for evaluating their reflections on **Investigation 7.**

Teacher Review

Use this section to reflect on and review the investigation. Keep in mind that your notes here are likely to be especially helpful when you teach this investigation again. Questions listed here are examples only.

Student Achievement

What evidence do you have that all students have met the science content objectives?

_____

Are there any students who need more help in reaching these objectives? If so, how can you provide this? _____

_____

What evidence do you have that all students have demonstrated their understanding of the inquiry processes? _____

_____

Which of these inquiry objectives do your students need to improve upon in future investigations? _____

_____

What evidence do the journal entries contain about what your students learned from this investigation? _____

_____

Planning

How well did this investigation fit into your class time? _____

_____

What changes can you make to improve your planning next time? _____

_____

Guiding and Facilitating Learning

How well did you focus and support inquiry while interacting with students?

_____

What changes can you make to improve classroom management for the next investigation or the next time you teach this investigation? _____

_____

How successful were you in encouraging all students to participate fully in science learning?

How did you encourage and model the skills values, and attitudes of scientific inquiry?

How did you nurture collaboration among students?

Materials and Resources

What challenges did you encounter obtaining or using materials and/or resources needed for the activity?

What changes can you make to better obtain and better manage materials and resources next time?

Student Evaluation

Describe how you evaluated student progress. What worked well? What needs to be improved?

How will you adapt your evaluation methods for next time?

Describe how you guided students in self-assessment.

Self Evaluation

How would you rate your teaching of this investigation?

What advice would you give to a colleague who is planning to teach this investigation?

Reflecting

Reflecting

Evidence for Ideas

Back to the Beginning

You have been investigating Our Dynamic Planet in many ways. How have your ideas changed since the beginning of the investigation? Look at the following questions and write down your ideas in your journal:

- What are volcanoes and why do they occur where they do?
- What are earthquakes and what causes them?
- How are earthquakes and volcanoes related?
- How do mountains form?

How has your thinking about earthquakes, volcanoes, and mountains changed?

Thinking about the Earth System

At the end of each investigation, you thought about how your findings connected with the Earth system. Consider what you have learned about the Earth system. Refer to the *Earth System Connection* sheet that you have been building up throughout this module.

- What connections between Dynamic Planet and the Earth system have you been able to find?

Thinking about Scientific Inquiry

You have used inquiry processes throughout the module. Review the investigations you have done and the inquiry processes you have used.

- What scientific inquiry processes did you use?
- How did scientific inquiry processes help you learn about the Dynamic Planet?

A New Beginning!

Not so much an ending as a new beginning!

This investigation into *Our Dynamic Planet* is now completed. However, this is not the end of the story. You will see the importance of Earth's dynamic events where you live and everywhere you travel. Be alert for opportunities to observe the importance of *Our Dynamic Planet* and add to your understanding.

Reflecting

This is the point at which your students review what they have learned throughout the module. This review is very important. Allow students time to work on this in a thoughtful way.

Back to the Beginning

These four questions were used as a pre-assessment (see **Blackline Master** *Our Dynamic Planet* P.1). Encourage students to complete this final review without looking at their journal entries from the beginning of the module. Their initial entries may influence their responses.

When students have completed their writing, encourage them to revisit their initial answers from the pre-assessment. Compare their writings at the end of the unit to their writings at the beginning. It is important that students not be left with the impression that they now know all there is to know about natural hazards associated with earthquakes and volcanoes. Emphasize that learning is a continuous process throughout their lives, and that practicing scientists themselves are still faced with a host of uncertainties and unanswered questions about our dynamic planet.

Assessment Opportunity

Comparisons between students' initial answers to these questions (in the pre-assessment at the beginning of the module) and those they are now able to give provide valuable data for assessment.

Thinking about the Earth System

Now that your students are at the end of this module, ask them to make connections between the geosphere and the other three parts of the Earth System. You may want to do this through a concept map. This is an opportunity for you to gauge how well students have developed their understanding of the Earth System for assessment purposes.

Thinking about Scientific Inquiry

To help students understand the relevance of these processes to their lives, ask them to think of everyday examples of when they use these processes (finding out where a misplaced book has gone; forming an opinion about a new TV show; winning an argument).

NOTES

Investigating Earth Systems – Investigating Our Dynamic Planet

Appendices

Investigating Our Dynamic Planet
Alternative End-of-Module Assessment

Part A: Matching
Write the letter of the term from Column B that best matches the description in Column A.

Column A	Column B
1. Bending of a wave due to changes in its velocity.	A. Computer model
2. Structure a scientist builds to represent something else.	B. Conceptual model
3. Fastest kind of seismic wave.	C. Convection
4. Model a scientist constructs in his or her mind.	D. P wave
5. Process that consumes ocean crust where two plates meet.	E. Physical model
6. Seismic wave that cannot travel through liquids.	F. R wave
	G. Refraction
	H. S wave
	I. Subduction
	J. Trench

Part B: Multiple Choice
Provide the letter of the choice that best answers the questions or completes the statement.

7. The oceanic crust is _____ than the continental crust.
   A. thicker
   B. thinner
   C. less dense
   D. smaller in area

8. The Earth's crust and mantle:
   A. have the same chemical composition
   B. have the same thickness
   C. have different densities
   D. have no boundary between them

9. The part of the Earth that is made of high-temperature liquid iron is called the:
   A. outer core
   B. inner core
   C. mantle
   D. asthenosphere
   E. lithosphere

10. Mantle _____ plays an important role in moving Earth's lithospheric plates:
   A. conduction
   B. convection
   C. refraction
   D. liquefaction

11. New crust is formed and spreads apart at:
    A. transform fault boundaries
    B. convergent plate boundaries
    C. subduction zones
    D. divergent plate boundaries

12. One plate slides past another horizontally at:
    A. transform fault boundaries
    B. convergent plate boundaries
    C. subduction zones
    D. divergent plate boundaries

13. The _____ of an earthquake is the point on the Earth's surface directly above the _____:
    A. epicenter / focus
    B. focus / epicenter
    C. refraction / epicenter
    D. refraction / focus

14. An earthquake on the sea floor can produce:
    A. Magma
    B. Subduction
    C. Tsunamis
    D. Tornadoes

15. What name did Alfred Wegener give to his theory of horizontal movement of the Earth's crust?
    A. Continental drift
    B. Isostasy
    C. Plate tectonics
    D. Sea-floor spreading

16. What name did Wegener give to his proposed single supercontinent?
    A. Eurasia
    B. Gondwanaland
    C. Panthalassa
    D. Pangea

17. Which of the following kinds of evidence did Wegener use to support his theory?
    A. Distributions of fossil plants and animals in the ocean
    B. Geographic fit of the continents
    C. Pattern of earthquakes and volcanoes along the "Ring of Fire"
    D. Presence of a large volcanic mountain range which he called the mid-ocean ridge
    E. All of the above

18. Why did most scientists of the 1920s reject Wegener's theory?
    A. The concentration of continents in the Northern Hemisphere
    B. Lack of a mechanism for continents to plow through oceanic crust
    C. The Earth was thought to be too young for such movements
    D. Wegener was not a geologist by training and so his ideas were ignored

19. The San Andreas Fault in southern California is an example of a boundary between:
    A. two plates that are colliding
    B. two plates that are moving apart
    C. two plates that are sliding past one another
    D. continental lithosphere and oceanic lithosphere

20. According to plate tectonics, new lithosphere is added to plates at a boundary between:
    A. two plates that are no longer moving
    B. two plates that are moving apart
    C. two plates that are sliding past one another
    D. continental lithosphere and oceanic lithosphere

21. What is the term used for molten rock within the Earth?
    A. granite
    B. pumice
    C. lava
    D. magma

22. The Earth's lithospheric plates move at speeds that average:
    A. 5 meters per year
    B. 5 centimeters per year
    C. 50 meters per year
    D. 5 kilometers year

23. Most earthquakes occur:
    A. in the middle of plates
    B. at the poles
    C. very deep within the Earth (>900 km deep)
    D. along plate boundaries

Part C: Essay

24. Refer to the diagram below to answer the following questions:

a) Using arrows, label the following features on the diagram: trench, oceanic crust, continental crust, volcanic arc, boundary between lithosphere and asthenosphere.

b) Explain why many earthquakes and volcanoes occur at the boundary depicted in the diagram below.

Answers

24.

A.

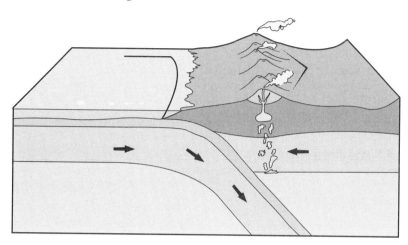

B. The boundary shown in the diagram is a subduction zone. At this kind of plate boundary, two lithospheric plates move toward one another and an oceanic lithospheric plate is forced below another plate. This oceanic plate is driven down into the mantle in a process called subduction. The interaction of these two plates generates forces that build up in the lithosphere. These forces can cause earthquakes, which are the plate's way of relieving the built-up stress. At a certain depth the subducting plate heats up enough to give off water or maybe even melt. This causes volcanoes to occur and explains why volcanoes are common near subduction zones.

Investigating Earth Systems Assessment Tools

Assessing the Student *IES* Journal

- Journal Entry-Evaluation Sheet
- Journal Entry-Checklist
- Key Question Evaluation Sheet
- Investigation Journal Entry-Evaluation Sheet
- Review and Reflect Journal Entry-Evaluation Sheet

Assisting Students with Self Evaluation

- Group Participation Evaluation Sheet I
- Group Participation Evaluation Sheet II

Assessing the Final Investigation

- Final Investigation Evaluation Sheet
- Student Presentation Evaluation Form

References

- Doran, R., Chan, F., and Tamir, P. (1998). *Science Educator's Guide to Assessment.*
- Leonard, W.H., and Penick, J.E. (1998). *Biology – A Community Context.* South-Western Educational Publishing. Cincinnati, Ohio.

Journal Entry-Evaluation Sheet

Name: _____ Date: _____ Module: _____

Explanation: The journal is an important component of each *IES* module. In using the journal as you investigate Earth science questions, you are mirroring what scientists do. The criteria, along with others that your teacher may add, will be used to evaluate the quality of your journal entries. Use these criteria, along with instructions within investigations, as a guide.

Criteria

1. Entry Made
       1   2   3   4   5   6   7   8   9   10   _____
Blank   Nominal   Above average Thorough

2. Detail
       1   2   3   4   5   6   7   8   9   10   _____
Few dates   Half the time   Most days   Daily
Little detail   Some detail   Good detail   Excellent detail

3. Clarity
       1   2   3   4   5   6   7   8   9   10   _____
Vague   Becoming clearer   Clearly expressed
Disorganized   well organized

4. Data Collection/Analysis
       1   2   3   4   5   6   7   8   9   10   _____
Data collected   Data collected,   Data collected
Not analyzed   some analyzed   and analyzed

5. Originality
       1   2   3   4   5   6   7   8   9   10   _____
Little evidence   Some evidence   Strong evidence
of originality   of originality   of originality

6. Reasoning/Higher-Order Thinking
       1   2   3   4   5   6   7   8   9   10   _____
Little evidence   Some evidence   Strong evidence
of thoughtfulness   of thoughtfulness   of thoughtfulness

7. Other
       1   2   3   4   5   6   7   8   9   10   _____

8. Other
       1   2   3   4   5   6   7   8   9   10   _____

Journal Entry-Checklist

Name: _____ Date: _____ Module: _____

Explanation: The journal is an important component of each *IES* module. In using the journal as you investigate Earth science questions, you are mirroring what scientists do. The criteria, along with others that your teacher may add, will be used to evaluate the quality of your journal entries. Use these criteria, along with instructions within investigations, as a guide.

Criteria

1. Makes entries _____

2. Provides dates and details _____

3. Entry is clear and organized _____

4. Shows data collected _____

5. Analyzes data collected _____

6. Shows originality in presentation _____

7. Shows evidence of higher-order thinking _____

8. Other _____

9. Other _____

Total Earned _____

Total Possible _____

Comments:

Key Question Evaluation Sheet

Name: _____ Date: _____ Module: _____

	No Entry		Fair		Strong
Shows evidence of prior knowledge	0	1	2	3	4
Reflects discussion with classmates	0	1	2	3	4

Additional Comments

Key Question Evaluation Sheet

Name: _____ Date: _____ Module: _____

	No Entry		Fair		Strong
Shows evidence of prior knowledge	0	1	2	3	4
Reflects discussion with classmates	0	1	2	3	4

Additional Comments

Key Question Evaluation Sheet

Name: _____ Date: _____ Module: _____

	No Entry		Fair		Strong
Shows evidence of prior knowledge	0	1	2	3	4
Reflects discussion with classmates	0	1	2	3	4

Additional Comments

Investigation Journal Entry-Evaluation Sheet

Name: _____ Date: _____ Module: _____

Criteria

1. Completeness of written investigation
   | 1 | 2 | 3 | 4 | 5 | 6 | 7 | 8 | 9 | 10 | _____ |

   Blank   Incomplete   Thorough

2. Participation in investigations
   | 1 | 2 | 3 | 4 | 5 | 6 | 7 | 8 | 9 | 10 | _____ |

   None or little;   Needs minimal guidance,   Leads, is inquisitive,
   unable to guide   sometimes helping others   persistent, focused
   self

3. Skills attained
   | 1 | 2 | 3 | 4 | 5 | 6 | 7 | 8 | 9 | 10 | _____ |

   Few skills   Tends to use some   High degree of
   evident   appropriate skills   appropriate skills used

4. Investigation Design
   | 1 | 2 | 3 | 4 | 5 | 6 | 7 | 8 | 9 | 10 | _____ |

   Variables not   Sometimes   Considers variables
   considered   considers variables,   Sound rationale for
   techniques   uses logical techniques   techniques
   illogical

5. Conceptual understanding of content
   | 1 | 2 | 3 | 4 | 5 | 6 | 7 | 8 | 9 | 10 | _____ |

   No evidence   Approaches understanding   Exceeds expectations
   of understanding   of most concepts   for content attainment

6. Ability to explain/discuss inquiry
   | 1 | 2 | 3 | 4 | 5 | 6 | 7 | 8 | 9 | 10 | _____ |

   Unable to   Some ability to   Uses scientific reasoning
   articulate   explain/discuss   to explain any
   scientific thought   the inquiry   aspect of the inquiry

7. Other
   | 1 | 2 | 3 | 4 | 5 | 6 | 7 | 8 | 9 | 10 | _____ |

8. Other
   | 1 | 2 | 3 | 4 | 5 | 6 | 7 | 8 | 9 | 10 | _____ |

Review and Reflect Journal Entry-Evaluation Sheet

Name: _____ Date: _____ Module: _____

Criteria	Blank		Fair		Excellent	
Thoroughness of answers	0	1	2	3	4	5
Content of answers	0	1	2	3	4	5
Other	0	1	2	3	4	5

Review and Reflect Journal Entry-Evaluation Sheet

Name: _____ Date: _____ Module: _____

Criteria	Blank		Fair		Excellent	
Thoroughness of answers	0	1	2	3	4	5
Content of answers	0	1	2	3	4	5
Other	0	1	2	3	4	5

Review and Reflect Journal Entry-Evaluation Sheet

Name: _____ Date: _____ Module: _____

Criteria	Blank		Fair		Excellent	
Thoroughness of answers	0	1	2	3	4	5
Content of answers	0	1	2	3	4	5
Other	0	1	2	3	4	5

Group Participation Evaluation Sheet I

Key:
4 = Worked on his/her part and assisted others
3 = Worked on his/her part
2 = Worked on part less than half the time
1 = Interfered with the work of others
0 = No work

My name is _____ .   I give myself a _____

The other people in my group are: I give each person:

A. _____   _____

B. _____   _____

C. _____   _____

D. _____   _____

Key:
4 = Worked on his/her part and assisted others
3 = Worked on his/her part
2 = Worked on part less than half the time
1 = Interfered with the work of others
0 = No work

My name is _____ .

The other people in my group are:

A. _____

B. _____

C. _____

D. _____

Group Participation Evaluation Sheet II

Name: _____ Date: _____ Module: _____

Key:
Highest rating _____
Lowest rating _____

1. In the chart, rate each person in your group, including yourself.

	Names of Group Members				
Quality of Work					
Quantity of Work					
Cooperativeness					
Other Comments _____					

2. What went well in your investigation?

3. If you could repeat the investigation, how would you change it?

Final Investigation Evaluation Sheet

Alerting students

Before your students begin the final investigation, they must understand what is expected of them and how they will be evaluated on their performance. Review the task thoroughly, setting time guidelines and parameters (whom they may work with, what materials they can use, etc.). Spell out the evaluation criteria for each level of proficiency shown below. Use three categories for a 3-point scale (Achieved, Approaching, Attempting). If you prefer a 5-point scale, add the final two categories.

Name: _____ Date: _____ Module: _____

	Understanding of concepts and inquiry	Use of evidence to explain and support results	Communication of ideas	Thoroughness of work
Exceeding proficiency 5	Demonstrates complete and unambiguous understanding of the problem and inquiry processes used.	Uses all evidence from inquiry that is factually relevant, accurate, and consistent with explanations offered.	Communicates ideas clearly and in a compelling and elegant manner to the intended audience.	Goes beyond all deliverables agreed upon for the project and has extended the data collection and analysis.
Achieved proficiency 4	Demonstrates fairly complete and reasonably clear understanding of the problem and inquiry processes used.	Uses the major evidence from inquiry that is relevant and consistent with explanations offered.	Communicates ideas clearly and coherently to the intended audience.	Includes all of the deliverables agreed upon for the project.
Approaching proficiency 3	Demonstrates general, yet somewhat limited understanding of the problem and inquiry processes used.	Uses evidence from inquiry to support explanations but may mix fact with opinion, omit significant evidence, or use evidence that is not totally accurate.	Completes the task satisfactorily but communication of ideas is incomplete, muddled, or unclear.	Work largely complete but missing one of the deliverables agreed upon for the project.
Attempting proficiency 2	Demonstrates only a very general understanding of the problem and inquiry processes used.	Uses generalities or opinion more than evidence from inquiry to support explanations.	Communication of ideas is difficult to understand or unclear.	Work missing several of the deliverables agreed upon for the project.
Non-proficient 1	Demonstrates vague or little understanding of the problem and inquiry processes used.	Uses limited evidence to support explanations or does not attempt to support explanations.	Communication of ideas is brief, vague, and/or not understandable.	Work largely incomplete; missing many of the deliverables agreed upon for the project.

Student Presentation Evaluation Form

Student Name_____ Date_____

Topic_____

	Excellent	Fair		Poor
Quality of ideas	4	3	2	1
Ability to answer questions	4	3	2	1
Overall comprehension	4	3	2	1

COMMENTS_____
_____
_____
_____
_____

Student Presentation Evaluation Form

Student Name_____ Date_____

Topic_____

	Excellent	Fair		Poor
Quality of ideas	4	3	2	1
Ability to answer questions	4	3	2	1
Overall comprehension	4	3	2	1

COMMENTS_____
_____
_____
_____
_____

Questions about Our Dynamic Planet

- **What are volcanoes and why do they occur where they do?**

- **What are earthquakes and what causes them?**

- **How are earthquakes and volcanoes related?**

- **How do mountains form?**

Use with *Our Dynamic Planet* Pre-assessment.

Student Journal Cover Sheet
Investigating Our Dynamic Planet

Name: _____

Group Members:

1. _____

2. _____

3. _____

4. _____

Teacher: _____

Class: _____

Dates of Investigation:

Start _____ Complete _____

Keep this journal with you at all times during your study of
Investigating Our Dynamic Planet.

Use with *Our Dynamic Planet* Pre-assessment.

Name: _____

Earth System Connection Sheet

When you finish an investigation, use this sheet to record any links you can make with the Earth system. By the end of the module you should have as complete a diagram as possible.

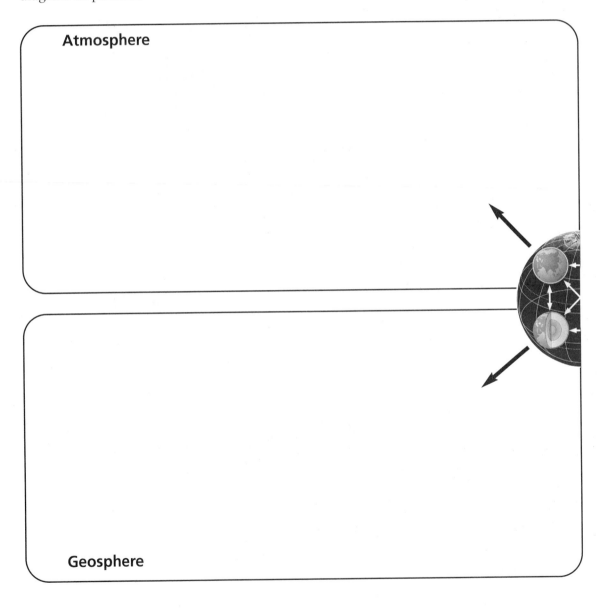

Use with *Our Dynamic Planet* Introducing the Earth System.

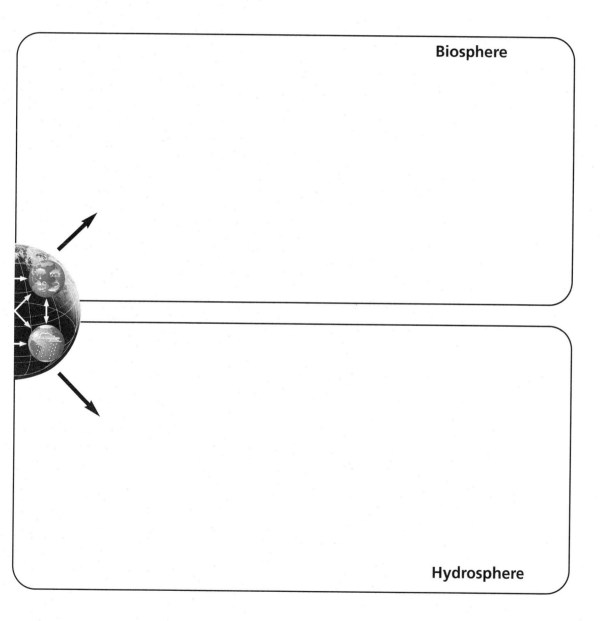

Biosphere

Hydrosphere

Inquiry Processes

 • Explore questions to answer by inquiry.

 • Design an investigation.

 • Conduct an investigation.

 • Collect and review data using tools.

 • Use evidence to develop ideas.

 • Consider evidence for explanations.

 • Seek alternative explanations.

 • Show evidence and reasons to others.

 • Use mathematics for science inquiry.

Use with *Our Dynamic Planet* Introducing the Earth System.

What are the Contents of the Mystery Bag?

WHAT ARE THE CONTENTS OF THE MYSTERY BAG?	
SMELL **Model** **Evidence**	**HEARING** **Model** **Evidence**
TOUCH **Model** **Evidence**	**FURTHER TESTS**

Use with *Our Dynamic Planet* Investigation 1: Gathering Evidence and Modeling

Class Data Table

Class Data – Measuring Speed of Water Waves

Group Number	Average Travel Time Seconds	Average Wave Speed (centimeters per second)
		1
		2
		3
		4
		5
Totals		
Class Average		

Use with *Our Dynamic Planet* Investigation 2: The Interior of the Earth

Modeling Wave Refraction

boundary line

20°

1.5 m

Refraction of Earthquake Waves

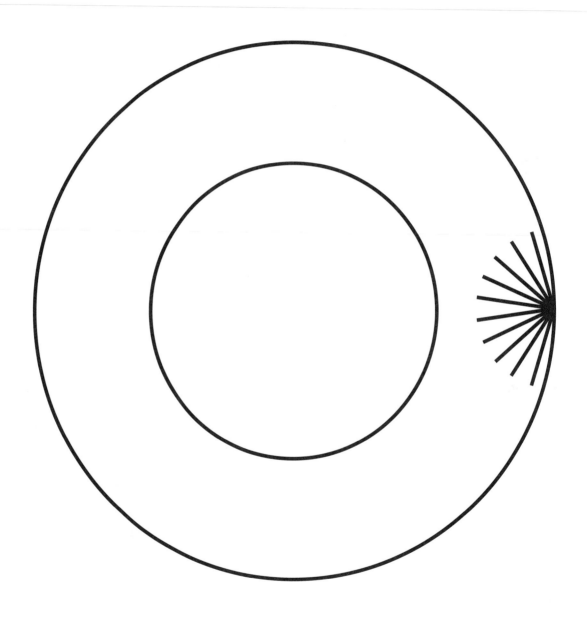

Use with *Our Dynamic Planet* Investigation 2: The Interior of the Earth

Seismic Wave Refraction and the Earth's Interior Structure

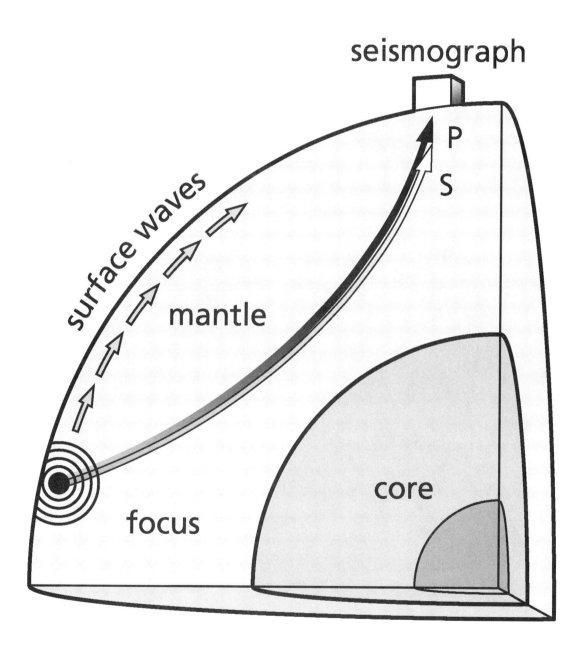

Use with *Our Dynamic Planet* Investigation 2: The Interior of the Earth

P wave Shadow Zone

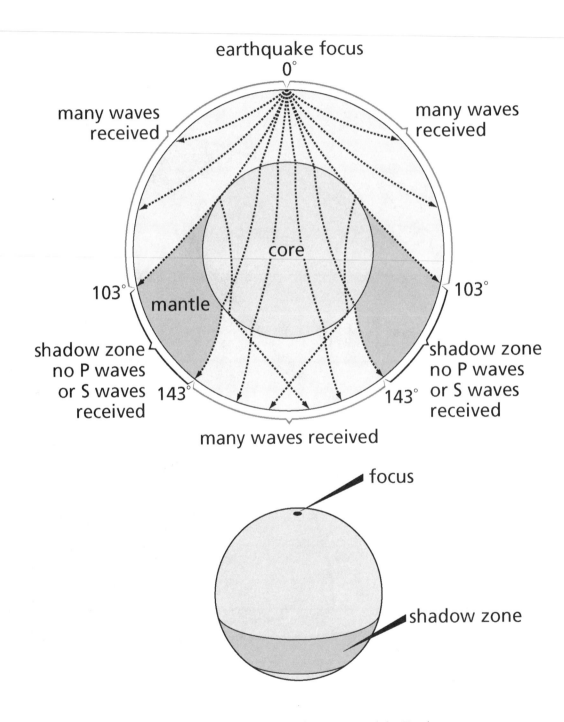

Use with *Our Dynamic Planet* Investigation 2: The Interior of the Earth

Earth's Interior Structure

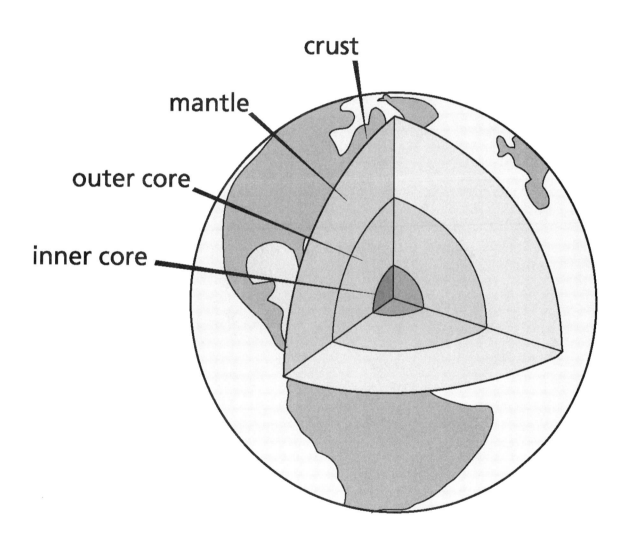

Experimental Setup

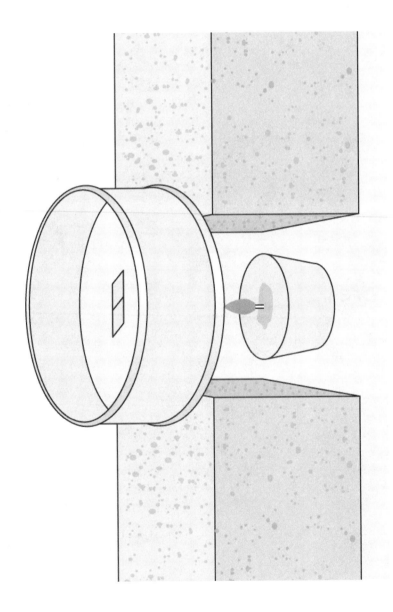

Use with *Our Dynamic Planet* Investigation 3: Forces that Cause Earth Movements

Convection Cells

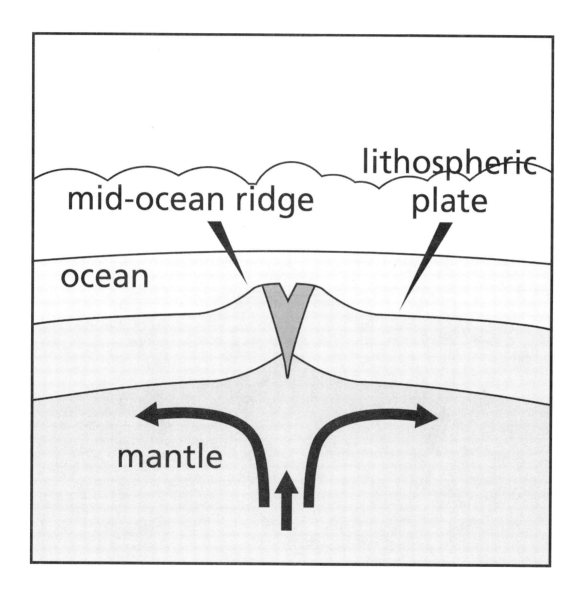

Mantle Convection

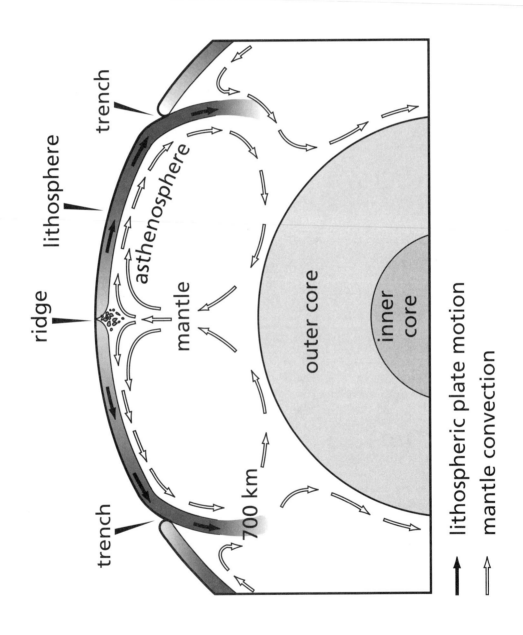

Use with *Our Dynamic Planet* Investigation 3: Forces that Cause Earth Movements

Major Lithospheric Plates and Kinds of Plate Boundaries

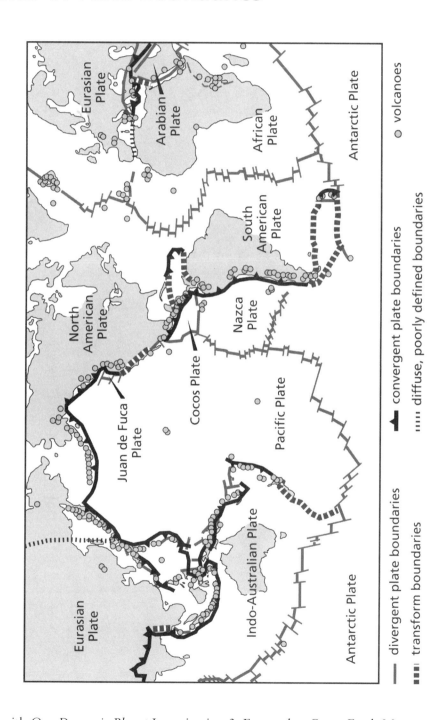

Use with *Our Dynamic Planet* Investigation 3: Forces that Cause Earth Movements

Sea-Floor Spreading

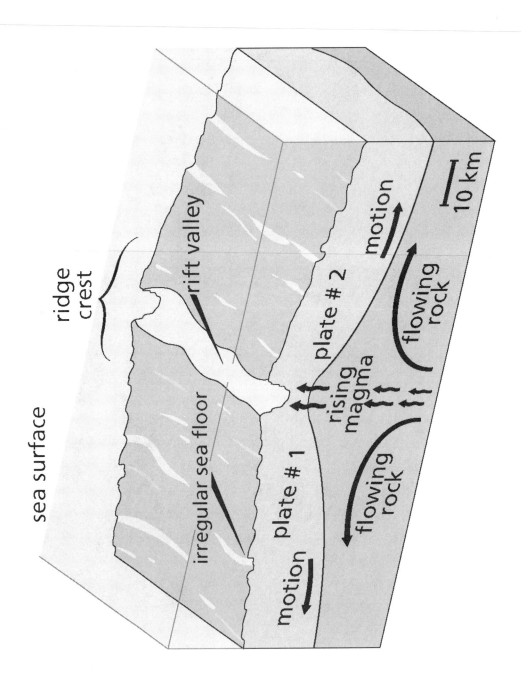

Experimental Setup – Investigation 4. Part A

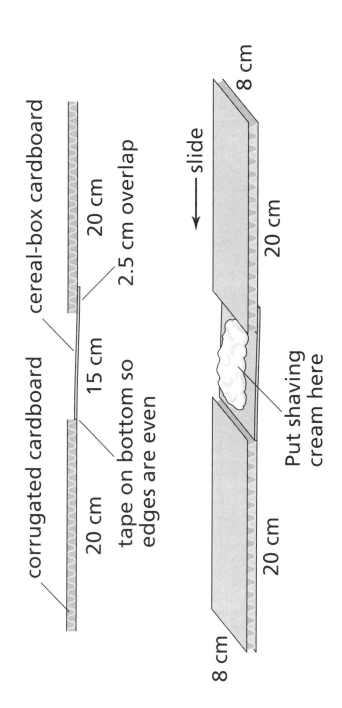

corrugated cardboard

cereal-box cardboard

20 cm

15 cm

20 cm

2.5 cm overlap

tape on bottom so edges are even

← slide

8 cm

8 cm

20 cm

20 cm

8 cm

Put shaving cream here

Use with *Our Dynamic Planet* Investigation 4: The Movement of the Earth's Lithospheric Plates

Relative Motion of Major Lithospheric Plates

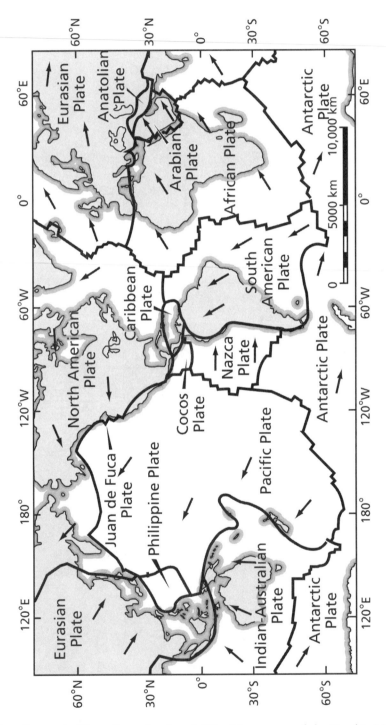

Use with *Our Dynamic Planet* Investigation 4: The Movement of the Earth's Lithospheric Plates

Lithosphere

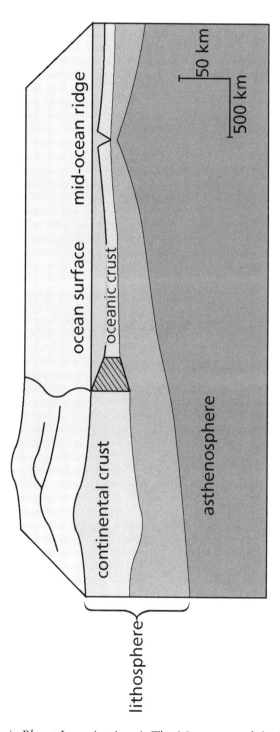

Types of Plate Convergence

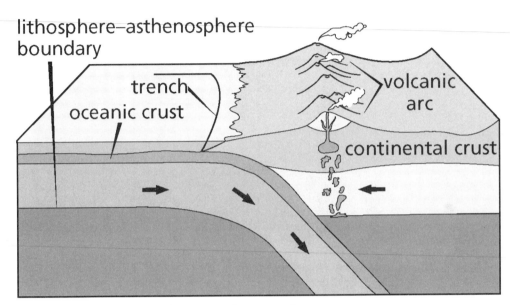

oceanic–continental convergence

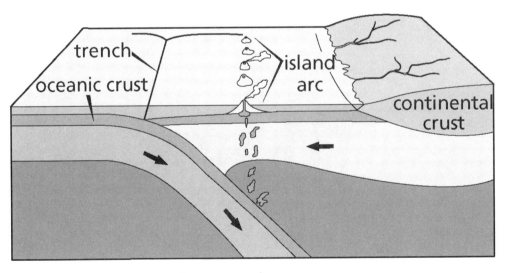

oceanic–oceanic convergence

Use with *Our Dynamic Planet* Investigation 4: The Movement of the Earth's Lithospheric Plates

Cross Section of Continent-Continent Convergence

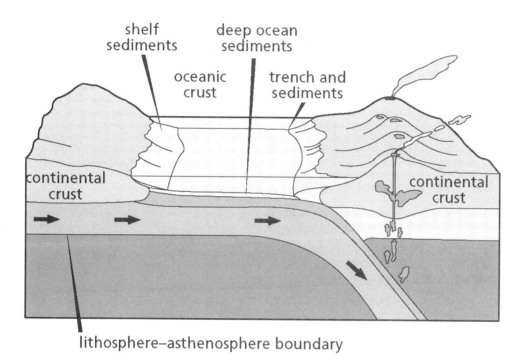

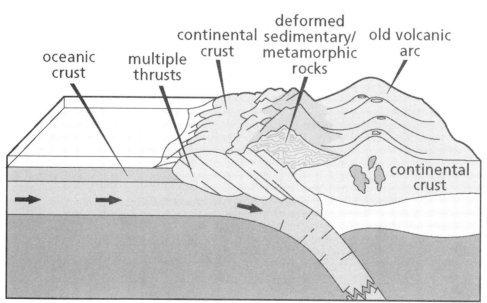

Use with *Our Dynamic Planet* Investigation 4: The Movement of the Earth's Lithospheric Plates

Collision of the Indian and Asian Plates

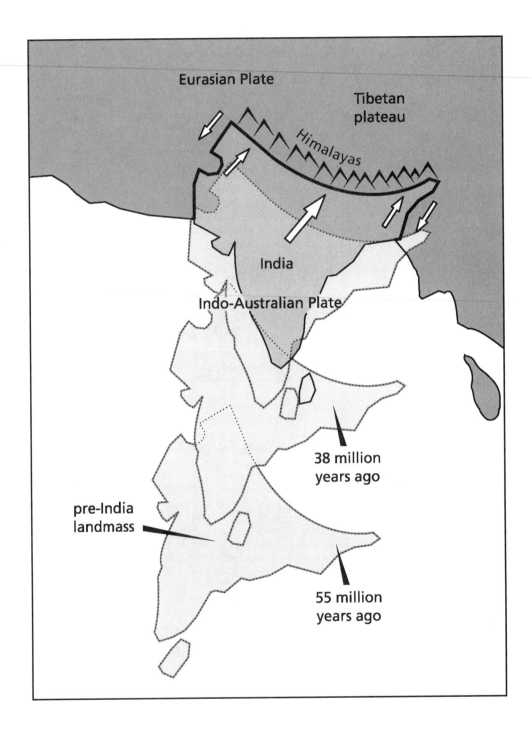

Eurasian Plate

Tibetan plateau

Himalayas

India

Indo-Australian Plate

38 million years ago

pre-India landmass

55 million years ago

Use with *Our Dynamic Planet* Investigation 4: The Movement of the Earth's Lithospheric Plates

World Map

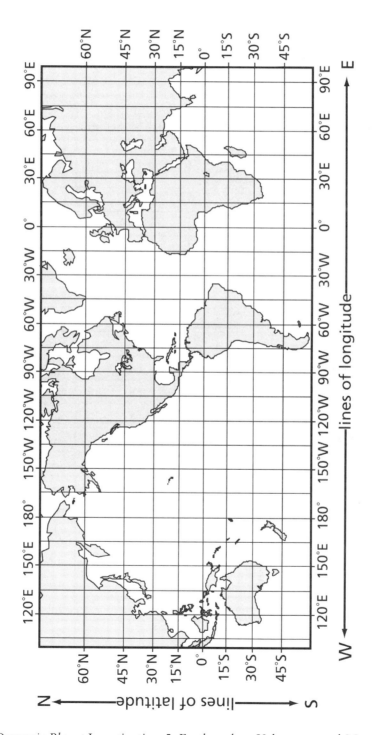

Use with *Our Dynamic Planet* Investigation 5: Earthquakes, Volcanoes, and Mountains

Seismograph Station Results for Five Days

Table 1: Subset of Seismograph Station Results for One Week

Latitude	Longitude	Depth (kilometers)	Magnitude (Richter Scale)	Occurrence Region
47°N	151°E	141	5.2	Kuril Islands
28°S	178°W	155	5.0	Kermadec Islands
30°N	52°E	33	4.2	Iran
36°N	140°E	69	4.7	Honshu, Japan
34°N	103°E	33	4.3	Gansu, China
40°S	177°E	27	4.8	New Zealand
0°N	36°E	10	4.6	Kenya, Africa
38°N	21°E	33	4.6	Ionian Sea
16°N	47°W	10	4.7	N. Mid-Atlantic Ridge
6°S	147°E	100	4.4	New Guinea
55°N	164°W	150	4.5	Unimak Island, Alaska
24°S	67°W	176	4.1	Argentina
13°N	91°W	33	4.2	Guatemala coast
4°N	76°W	171	5.6	Colombia
40°N	125°W	2	4.5	N. California coast
5°S	102°E	33	4.4	S. Sumatra, Indonesia
44°S	16°W	10	4.6	S. Mid-Atlantic Ridge
51°N	179°E	33	4.4	Aleutian Islands
15°S	71°W	150	4.2	Peru
49°N	128°W	10	4.7	Vancouver, Canada
35°N	103°E	33	4.3	Gansu, China

Use with *Our Dynamic Planet* Investigation 5: Earthquakes, Volcanoes, and Mountains

Table 1: Global Volcanic Activity Over One-Month Period

Table 2: Global Volcanic Activity Over One-Month Period

Latitude	Longitude	Location	Region
1°S	29°E	Nyamuragira	Congo, Eastern Africa
38°N	15°E	Stromboli	Aeolian Islands, Italy
37°N	15°E	Etna	Sicily, Italy
15°S	71°W	Sabancaya	Peru
0°	78°W	Guagua Pichincha	Ecuador
12°N	87°W	San Cristobal	Nicaragua
0°	91°W	Cerro Azul	Galapagos, Ecuador
19°N	103°W	Colima	Western Mexico
19°N	155°W	Kilauea	Hawaii, USA
56°N	161°E	Shiveluch	Kamchatka, Russia
54°N	159°E	Karymsky	Kamchatka, Russia
43°N	144°E	Akan	Hokkaido, Japan
39°N	141°E	Iwate	Honshu, Japan
42°N	140°E	Komaga-take	Hokkaido, Japan
1°S	101°E	Kerinci	Sumatra, Indonesia
4°S	145°E	Manam	Papua, New Guinea
5°S	148°E	Langila	Papua, New Guinea
15°S	167°E	Aoba	Vanuatu
16°N	62°W	Soufriere Hills	Montserrat, West Indies
12°N	86°W	Masaya	Nicaragua
37°N	25°W	Sete Cidades	Azores

Use with *Our Dynamic Planet* Investigation 5: Earthquakes, Volcanoes, and Mountains

Major World Mountain Chains

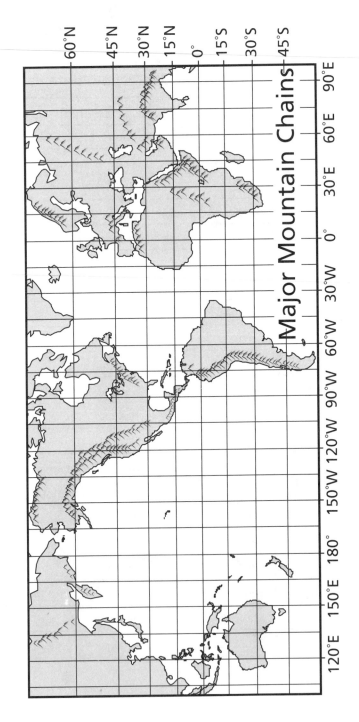

Use with *Our Dynamic Planet* Investigation 5: Earthquakes, Volcanoes, and Mountains

World Map of Continents and Continental Shelves

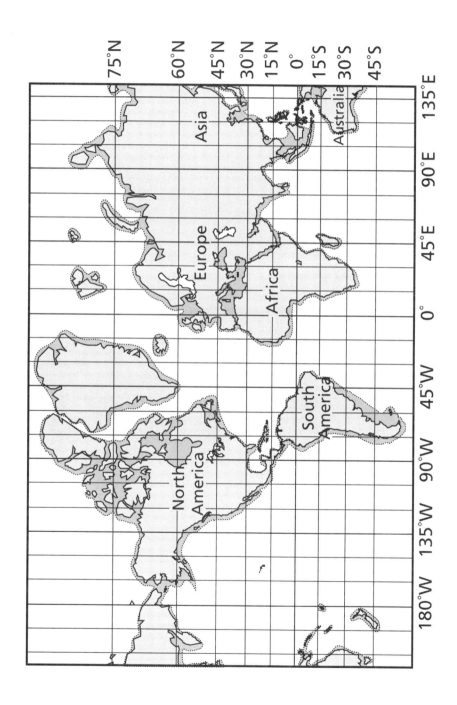

Use with *Our Dynamic Planet* Investigation 6: Earth's Moving Continents

Ancient Mountain Belts

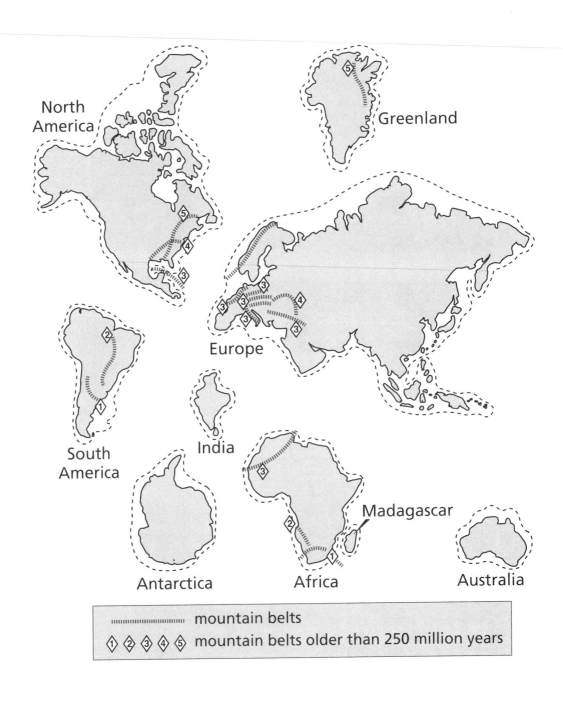

Use with *Our Dynamic Planet* Investigation 6: Earth's Moving Continents

Generalized Distribution of Fossils of Lystrosaurus, Glossopteris, Cynognathus, and Mesosaurus

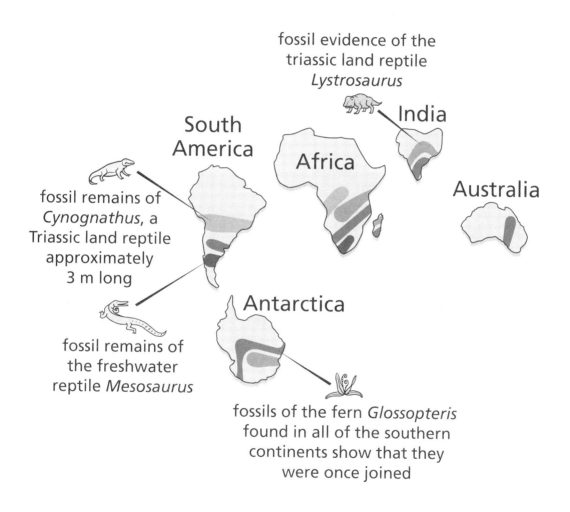

fossil evidence of the triassic land reptile *Lystrosaurus*

South America

Africa

India

Australia

fossil remains of *Cynognathus*, a Triassic land reptile approximately 3 m long

fossil remains of the freshwater reptile *Mesosaurus*

Antarctica

fossils of the fern *Glossopteris* found in all of the southern continents show that they were once joined

Use with *Our Dynamic Planet* Investigation 6: Earth's Moving Continents

Ice Sheet Distribution, 300 Million Years Ago

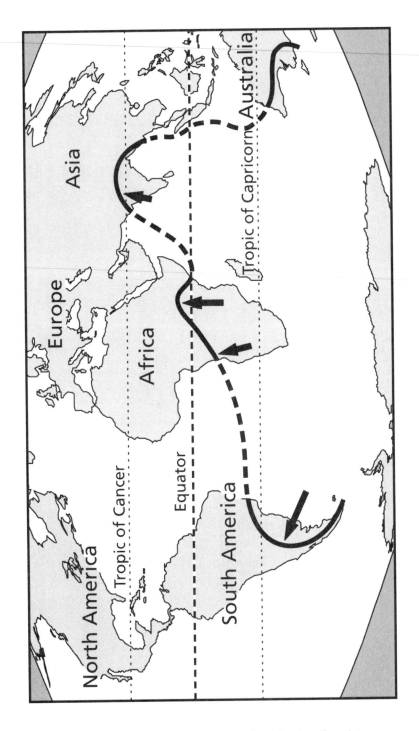

Use with *Our Dynamic Planet* Investigation 6: Earth's Moving Continents

Pangea

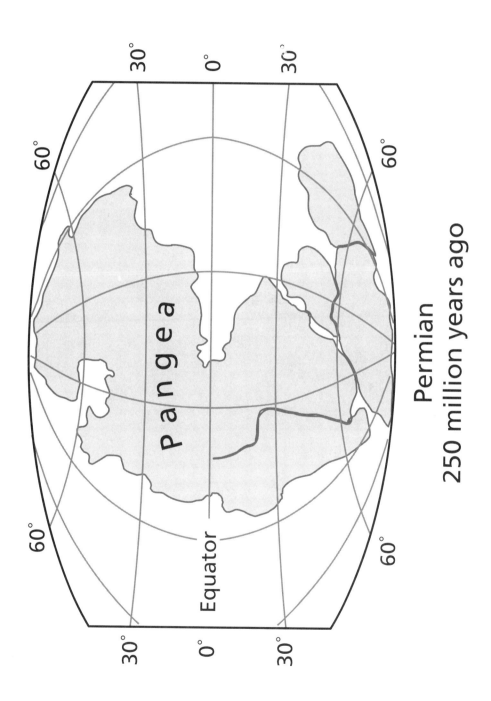

Pangea

Equator

Permian
250 million years ago

Use with *Our Dynamic Planet* Investigation 6: Earth's Moving Continents

Breakup of Pangea

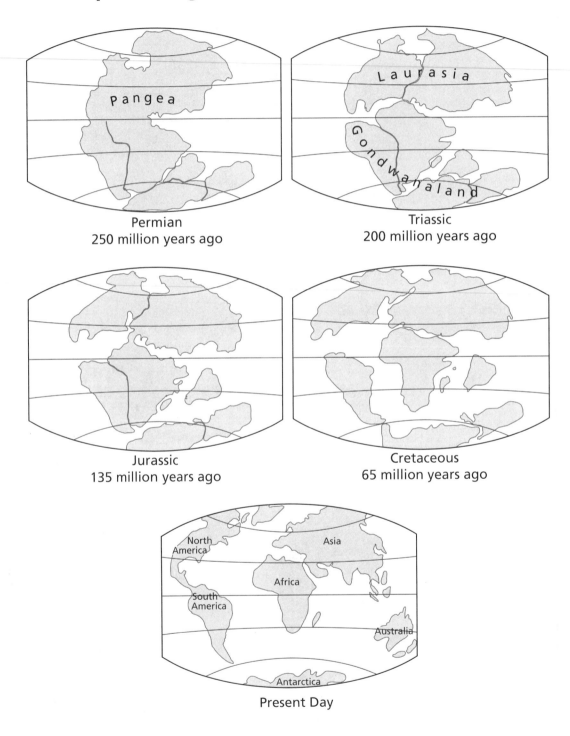

Permian
250 million years ago

Triassic
200 million years ago

Jurassic
135 million years ago

Cretaceous
65 million years ago

Present Day

Use with *Our Dynamic Planet* Investigation 6: Earth's Moving Continents

Table – Earthquake and Volcano Hazards

Event	Effect	Examples
Volcanoes	Eruption with lava flow	• lava streams burn all in their path
	Eruption with ash fall	• aircraft endangered, roofs collapse
	Lahar (mud flow)	• Large, fast-moving river of mud
	Pyroclastic flow	• hot mobile flow of volcanic material
	Lateral blast	• explosive wave knocks down all in its path
	Volcano collapses	• land drops away— homes threatened
	Volcanic gases	• volcanic pollution
Earthquakes	Ground motion	• buildings and bridges collapse
	Fault displacement	• roads crack, rail tracks split
	Fires	• ruptures in gas lines cause building fires
	Landslides	• shaking causes rock to slide downhill
	Liquefaction	• ground becomes like quicksand
	Failures of dams	• flooding

Use with *Our Dynamic Planet* Investigation 7: Natural Hazards and Our Dynamic Planet

Schematic Diagram – Explosive Volcanic Eruption

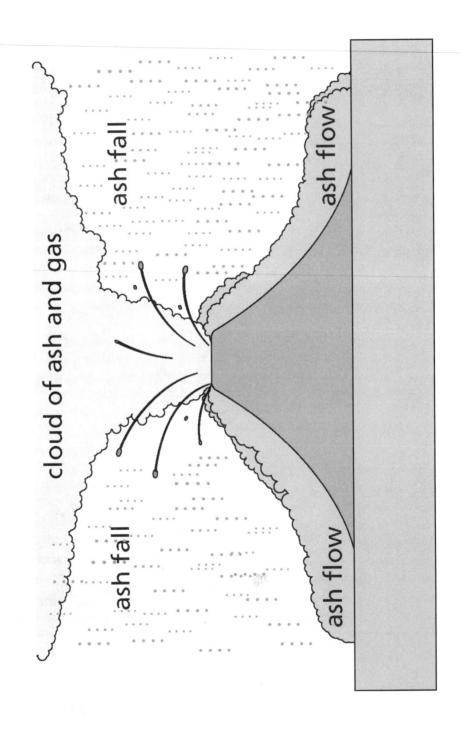

Use with *Our Dynamic Planet* Investigation 7: Natural Hazards and Our Dynamic Planet

NOTES

NOTES

Investigating Earth Systems – Investigating Our Dynamic Planet